Richard E. Call, Constantine S. Rafinesque

Ichthyologia Ohiensis

Natural history of the fishes inhabiting the river Ohio and its tributary

streams

Richard E. Call, Constantine S. Rafinesque

Ichthyologia Ohiensis
Natural history of the fishes inhabiting the river Ohio and its tributary streams

ISBN/EAN: 9783337302078

Printed in Europe, USA, Canada, Australia, Japan

Cover: Foto ©berggeist007 / pixelio.de

More available books at **www.hansebooks.com**

ICHTHYOLOGIA OHIENSIS,

OR

NATURAL HISTORY

OF

THE FISHES INHABITING THE

RIVER OHIO

AND ITS TRIBUTARY STREAMS,

Preceded by a physical description of the Ohio and its branches.

BY C. S. RAFINESQUE,

Professor of Botany and Natural History in Transylvania University, Author of the Analysis of Nature, &c. &c. Member of the Literary and Philosophical Society of New-York, the Historical Society of New-York, the Lyceum of Natural History of New York, the Academy of Natural Sciences of Philadelphia, the American Antiquarian Society, the Royal Institute of Natural Sciences of Naples, the Italian Society of Arts and Sciences, the Medical Societies of Lexington and Cincinnati, &c. &c.

The art of seeing well, or of noticing and distinguishing with accuracy the objects which we perceive, is a high faculty of the mind, unfolded in few individuals, and despised by those who can neither acquire it, nor appreciate its results.

LEXINGTON, KENTUCKY:

PRINTED FOR THE AUTHOR BY W. G. HUNT, (PRICE ONE DOLLAR.)

1820.

These Pages

and the Discoveries which they contain

in one of the principal Branches

of Natural History,

are respectfully Inscribed

by the Author;

To his fellow-labourers in the same field of Science:

Prof. SAMUEL L. MITCHILL, M. D.

who has described the Atlantic Fishes of New York,

and to

C. A. LE SUEUR,

who was the first to explore the Ichthyology of the

Great American Lakes, &c.

In Token

of Friendship, Respect, and Congratulation.

NATURAL HISTORY

OF THE FISHES OF THE OHIO RIVER AND ITS TRIBUTARY STREAMS,

BY C. S. RAFINESQUE,

Professor of Botany and Natural History in Transylvania University.

INTRODUCTION.

Nobody had ever paid any correct attention to the fishes of this beautiful river, nor indeed of the whole immense basin, which empties its water into the Mississippi, and hardly twelve species of them had ever been properly named and described, when in 1818 and 1819, I undertook the labour of collecting, observing, describing, and delineating those of the Ohio. I succeeded the first year in ascertaining nearly eighty species among them, and this year I added about twenty more, making altogether about one hundred species of fish, whereof nine tenths are new and undescribed.

Many of them have compelled me to establish new genera, since they could not properly be united with any former genus; and I could have increased their number, had I been inclined, as will be seen in the course of this ichthyology; but I have in many instances proposed sub-genera and sections instead of new genera. I sent last spring to Mr. Blainville of Paris, a short account of some of them, to be published in his Journal of Natural History, in a Tract named *Prodromus of seventy new genera of Animals and fifty new genera of Plants from North America,* and I now propose to publish a complete account of all the species I have discovered. I am confident that they do not include the whole number existing in the Ohio, much less in the Mississippi; but as they will offer a great proportion of them, and, as the additional species may be gradually described in supplements, I venture to introduce them to the acquaintance of the American and European naturalists; being confident that they will not be deemed an inconsiderable addition to our actual knowledge of the finny tribes. To the inhabitants of the western states, to those who feed daily upon them, their correct and scientific account ought to be peculiarly agreeable. I trust they will value the exertions through which I have been able to accomplish so much in so short a period of time, and I wish I could induce them to lend me their aid, in the succession of my studies of those animals, by communicating new facts, details, and rare species. I may assure them that their kind help shall be gratefully received and acknowledged.

The science of Ichthyology has lately received great additions in the United States. A few of the atlantic fishes had been formerly enumerated by Catesby, Kalm, Forster, Garden, Linnæus Schœpf, Castiglione, Bloch, Bosc, and Lacepede: but Dr. Samuel L. Mitchell has increased our knowledge, with about one hundred new species at once, in his two memoirs on the Fishes of New-York, the first published in 1814, in the Transactions of the Literary and Philosophical Society of New-York, and the second in the American Monthly Magazine in 1817. Mr. Lesueur was the first naturalist who visited Lake Erie and Lake Ontario, where he detected a great number of new species, which he has already begun to publish in the Journal of the Academy of Sciences of Philadelphia, and which he means to introduce in his General History of American Fishes, a work on the plan of Wilson's Ornithology, which he has long had in contemplation. And I have added thereto about forty new species, which I discovered in Lake Champlain, Lake George, the Chesapeake, the Hudson, near New-York, Philadelphia, the Atlantic, &c. and published in my *Precis des Decouvertes*, my Memoirs on Sturgeons, my decads and tracts in the American Monthly Magazine, the American Journal of Science, &c. besides three new fishes of the Ohio, published in the Journal of the Academy of Philadelphia.

Many other fishes of the United States have been partially

described by Bartram, Carver, Lewis and Clarke and other travellers. It is reasonable to suppose that several others have es-
caped their notice, and my discoveries in the Ohio prove this
assertion. I calculate that we know at present about five hundred species of North American fishes, while ten years ago we
hardly knew one hundred and twenty. Among that number a-
bout one half are fresh water fishes, and one fourth at least belong to the waters of the western states; but, although there are
fifty other species imperfectly known, I should not wander far
from reality if I should conjecture that, after all, we merely know
one third of the real numbers, when we consider that the whole
of the Mexican Provinces is a blank in Ichthyology, as well as
California, the North West Coast, the Northern Lakes, and all
the immense bason of the Missouri and Mississippi, except the
eastern branch of the Ohio: all those regions having never been
explored by any real naturalists. From those who are actually
surveying the river Missouri much may be expected; but I venture to foretell that many of the fishes of the Ohio will be found
common to the greatest part of the streams communicating
with it, and therefore throughout the Mississippi and Missouri,
whence the ichthyology of the Ohio, will be a pretty accurate
specimen of the swimming tribes of all the western waters;
while in Mexico, the North West Coast, and in the basin of
the St. Lawrence or even in the Floridian waters, a total difference of inhabitants may be detected: since I have already ascer
tained that out of one hundred species of Ohio fishes, there are
hardly two similar to those of the atlantic streams.

I have in contemplation to visit many other western streams
and lakes, where I have no doubt to reap many plentiful har-
vests of other new animals; meantime communications on the
fishes of every western stream are solicited from those, who
may be able and willing to furnish them.

It is probable that some of the fishes of the Mississippi
are anadromic or come annually from the gulf of Mexico to
spawn in that stream and its lower branches; but all the fishes
of the Ohio remain permanently in it, or at utmost travel down
the Mississippi during the winter, although the greatest proportion dwell during that season in the deep spots of the Ohio

This is proved by their early appearance at the same time in all the parts of the river and even as high as Pittsburgh. This happens even with the Sturgeons and Herrings of the Ohio, which are in other countries periodical fishes, travelling annually from the sea to the rivers in the spring, and from the rivers to the sea in the fall.

Fishes are very abundant in the Ohio, and are taken sometimes by thousands, with the seines: some of them are salted; but not so many as in the great lakes. In Pittsburgh, Cincinnati, Louisville, &c. fish always meets a good market; and sells often higher than meat; but at a distance from those towns you may buy the best fish at the rate of one or two cents the pound. It affords excellent food, and, if not equal to the best sea fish, it comes very near it, being much above the common river fish of Europe: the most delicate fishes are the Salmon-perch, the Bubbler, the Buffaloe-fish, the Sturgeons, the Crabshes, &c. It is not unusual to meet such fishes of the weight of thirty to one hundred pounds, and some monstrous ones are occasionally caught, of double that weight. The most usual manners of catching fish in the Ohio are, with seines or harpoons at night and in shallow water, with boats carrying a light, or with the hooks and lines, and even with baskets.

I am sorry to be compelled to delay the publication of my figures of all the fishes now described: these delineations shall appear at another period.

To facilitate the knowledge of the streams mentioned, I prefix a physical description of the Ohio and its principal branches.

Lexington, Kentucky, 15th November, 1819.

RIVER OHIO.

HEAD. It is formed by the junction of the rivers Alleghany and Monongahela, in Pennsylvania, at Pittsburgh, near the 40½ degree of north latitude. It is difficult to say which of them is the main branch or stream, the Alleghany being the longest and in the most direct course, while the Monongahela appears to be the largest at the junction, and to have similar waters.

DIRECTION. Although the Ohio is exceedingly crooked in its course, its general direction is south west and west south west: it assumes every other direction; but very seldom the opposite one, N. E.

MOUTH. It empties into the Mississippi, near the 37th degree of latitude, dividing the state of Kentucky from that of Illinois, which lies north.

CONNECTIONS. The Ohio is one of the principal branches of the Mississippi, and properly its great eastern branch. The two great western branches, the Arkansas, which is about 1800 English miles long, and the Red River, which measures about 1600 miles, exceed it in length, but not in size, nor in the number of tributary streams; nor in the extent of their basins. The northern branch or upper Mississippi is much inferior to it in all respects (it is only 775 miles long, and receives only seven large rivers,) although it has been mistaken for the main branch. The real main branch is the Missouri, which takes the name of Mississippi after its junction with the upper Mississippi. It flows 2700 English miles above that junction, receiving thirty-three rivers above 100 miles long, and 1300 miles below, receiving twelve such rivers, having a total course of 4000 miles and forty five large branches. It is yet undecided whether the Yellow Stone or the Western Missouri is the principal upper branch.

LENGTH. From Pittsburgh to the mouth, it is 500 geographic miles in a direct course (60 to a degree) and 960 in the regular course, equal to 1120 English miles, (of 69¼ to a degree;) but if the Monongahela be deemed the main upper branch, the whole course will be 1360 English miles, while if the Allegany be considered as such, the whole length of the Ohio will be found equal to 1405 such miles.

ADJACENCIES. It flows through Pennsylvania as far as Mill creek below Georgetown; it divides afterwards the state of Ohio, which lie on the right bank from Virginia; this state extends on the left bank as far as Sandy river, where Kentucky begins, and it occupies the remainder of the left bank, as far as the Mississippi. While the state of Ohio terminates on the north side at the Miami river: the state of Indiana follows as far as the Wabash river, and from thence the state of Illinois extends to the mouth.

PARTS. The Ohio is naturally divided into three parts, containing each two sections, the head branches Alleghany and Monongahela form the two sections of the first part. The second or upper part lies between their junctions and the falls, being divided into two sections by Letart's rapids; while the third or lower part includes the space below the falls, the first section of which terminates at the end of the narrow valley above Troy in Indiana, and the second which includes the broad and flat valleys reaches to the the mouth. The upper part of the river is the longest, being about seven hundred miles long.

BREADTH. At Pittsburgh the Ohio is about one quarter of a mile wide, above the falls and near the mouth it is over one mile: its average breadth may be reckoned at half a mile or rather two thousand five hundred feet.

DEPTH. Very variable according to places and times. The mean depth at low water may be reckoned at three feet, and at high water at about thirty feet. Average medium fifteen feet.

VELOCITY. The current of the Ohio is generally gentle, except at the falls and ripples. Its average at low water may be stated at two miles an hour and at high water at four miles an hour.

BULK. The quantity of waters flowing in the Ohio may be therefore calculated, upon a general medium of the above breadth, depth, and velocity, at about forty millions of cubic feet, during an hour at low water, and at more than eight hundred millions of such feet at high water. Average medium three hundred and eighty millions in an hour, nine thousand one hundred and twenty millions in a day, and more than three millions of millions of feet in one year.

WATERS. They are slightly turbid, and become much more so in the rises. At a low stage they are almost clear, and at all times very salubrious. The Monongahela has the same character, while the Alleghany is almost perfectly clear. The turbidity of the waters is produced by very fine particles of earthy matter dissolved in it, and which are not easily deposited, unless at high water, when mud and earth become mixed with them.

VALLEY. The Ohio flows in a narrow valley as far as Utica, a-
bove Louisville. This valley averages about one mile in breadth,
and about three hundred feet in depth, but in some parts it is
nearly five hundred teet deep. There are evident proofs that
the river has formerly filled it. The sides are formed by steep
cliffs and hills of sandstone as far as Vanceburg and the knobs
below the mouth of the Scioto; beyond which all the strata are
of limestone. Beyond those cliffs the country is broken, but
much raised above the bottom of the Ohio Valley. The river
meanders through it, leaving on each side, or only on one side,
a level tract of alluvial and deep soil, which are called *bottoms*
and were once the bed of the river. The cliffs correspond together, keeping at a equal distance, and every salient angle
or elbow has an opposite bend. Below Utica and as far as Otter creek below Salt river begins the site of an ancient Lake,
forming now a plain, which is about twenty-five miles long and
ten miles broad; the falls are situated in the middle of it: the
silver hills bound it to the west, the knobby hills to the east and
the barren hills to the south. Immediately below it are the
narrows of Otter creek, where the valley begins again; but is
not larger than at Pittsburgh, being hardly half a mile wide and
the river is less than one thousand feet across. They both expand gradually until they reach the rocky narrows above Troy,
where the valley, after being contracted to three fourths of a
mile, while the river is nearly half a mile broad, expands at
once into a low country or broad valley, (the river being often
one mile wide) which was formerly a second lake, extending
about one hundred miles to Cave-hill narrows, with a variable
breadth of four to twenty miles; only a few bluffs appearing occasionally on the banks, and the boundary hills being only one
hundred and fifty feet high on an average. At Cave-hill or
Cave in the rock, the river, from a mile broad, becomes at once
very narrow, and the hills come very near the banks on both
sides, forming a short narrows, below which stands another
plain, which was once a third Lake, about twelve miles long
and six miles wide: it ends at Grand Pierre creek, and the broad
narrows between the north and south bluffs. Here begins the
lowest part of the Ohio Valley, which grows wide gradually

and extends as far as the Mississippi, being from six to twenty miles wide and bounded by hills one hundred feet high on an average, and with very few stones.

BASIN. The basin of a river, must not be mistaken for its valley, since it includes the whole regions watered by the streams flowing into it. The basin of Ohio is very extensive, including the greater share of the states of Kentucky, Tennessee, Ohio, and Indiana, with parts of Pennsylvania, New-York, Virginia, Alabama and Illinois, and a small corner of North Carolina, Georgia and Mississippi, watering therefore twelve states of the Union. It occupies eight degrees of latitude from the thirty-fourth to the forty-second degrees, and about twenty-six degrees of longitude. Its whole surface includes at least half a million of square miles, and three hundred and twenty millions of square acres.

ISLANDS. The Ohio has a great many, about one hundred and thirty; they are commonly long and narrow. Some sand-bars, lying in the middle of the river, are gradually becoming islands; most of them are overflowed at the high waters. There are very few ancient islands, forming now insulated hills; I have detected however half a dozen, the first of which lies just below Pittsburgh on the right bank.

BARS. They are very common, are generally sand bars, and lie on one side or round the islands, very few stretch across the river: they produce ripples or a broken current. Some of them have hardly six inches of water, at the low stage of the river.

CHANNELS. The current of the Ohio is digging another bed, deeper than the actual one, which forms the real channel of navigation. It does not experience many changes; sometimes it happens to be very crooked, particularly near islands and bars. It generally follows and grazes the highest cliffs or banks, and sometimes becomes double round some islands.

BANKS. The actual banks are all alluvial and of a deep and rich soil, seldom quite sandy or muddy. There are in many bottoms a second and even a third bank, all very steep and from ten to forty feet high. The first bank is almost every where overflowed at high waters, the second never. The platforms behind the banks are sometimes lower than the edge of the bank. Many banks sink or are washed away in inundations, when the channel sets against them.

RAPIDS. Many ripples become rapids at low water, and all the rapid disappear at high water, even those called the falls, which lie below Louisville. They are several passages of the river between rocky islands, the waters flowing with great rapidity; but hardly ever pitching over, except on the Kentucky side of the falls, where at very low water there is a small fall of less than two feet. Their noise is heard at a great distance. A Canal will soon be cut on each side of them. Letart's rapids and the Hurricane rapids are the most dangerous after the falls, yet they are merely large rock ripples.

BAYOUS. They are narrow channels into which the waters flow at a certain stage of rise, forming temporary islands; they are not uncommon in the lower vallies, and are sometimes called cut offs; the longest lies below Evansville, forming occasionally a very large island opposite Hendersonville.

INUNDATIONS The Ohio is subject to periodical rises and to many adventitious ones. The highest happens in the spring, when the snow melts in the Alleghany mountains, and it has sometimes risen to fifty feet above the low water at some particular places, covering all the islands and bottoms of the first banks, and overflowing the towns built on those bottoms, such as Marietta, Shippingport, Lawrenceburgh, Shawneetown, &c. to the depth of ten feet or more. Another happens in the fall after the first rains; both subside pretty soon. Many others occur throughout the year, occasioned by rains. They are either general or partial, sudden or gradual; but during the months of July, August, and September the waters are very low, while in January and February, they are covered with floating ice and even frozen over in the northern and upper part. The over flowings do not rise so high in the lower valleys; but they expand more over the bottoms, often leaving behind pools and marshes.

PHENOMENA. Eddies and whirlpools are common, particularly at high waters; but not dangerous. A natural echo is heard throughout the narrow valley. Fogs are common dur-

B

ing the winter and spring in the valley, they collect in the morn-
ing and last until the sun dissipates them: they preserve the
valley from the chilling frosts, and render its climate milder
than that of the adjacent country. The prevailing winds are
westerly,and four times out of five a breeze blows up the stream,
following the meanders of the valley: it is a deviated branch
of the Mexican trade wind. Thunder storms are frequent in
summer, and hurricanes have sometimes happened. Waves
then rise high against the current and are dangerous. Inter-
mittent fevers are not uncommon in the fall near some low banks
and in the low bottoms; but the climate is otherwise very
healthy. Many springs are found along the banks and cliffs
and many more appear at low water.

SCENERY. All the banks, and cliffs, and nearly all the islands
are covered with trees, among which the *Platanus occidentalis*
(Sycamore,) the *Populus angulata*, (Cotton tree,) and the *Sa-
lix nigra* (Willow) are the most common and conspicuous.
The cliffs and islands offer every where very fine views and
prospects, and the cultivation increases those natural beau-
ties; this is very conspicuous near Cincinnati, Maysville, Pittsburgh, &c.

NAVIGATION. The River is navigated by Steam boats, Barges, Keel boats, Schooner barges, Rowing boats, Flat boats or
Arks, Skiffs, Pirogues, Rafts, &c. of which many thousand annually descend the stream. Those which ascend it again a-
mount annually to many hundred, among which there are already more than sixty Steam boats, averaging the burthen of
150 tons each. The ascent is effected, besides steam, by sailing,
poling, warping,and rowing, and is very tedious. The difficul-
ties of the navigation consist in bars, sunken rocks, rocky ledg-
es, snags or sunken logs, sawyers or moving snags, drifted logs,
planters or upright trees, falling trees, sinking banks, sudden
storms, rises and falls, drifting ice, rejecting currents, whirlpools, shallow water, ripples and rapids, &c. : but they are not
dangerous except at some particular stages of the waters. In
the spring rise the water is so deep that it may easily float ves-
sels of 500 tons, even over the falls. Many large ships were
built at Pittsburgh and Marietta, which safely reached the sea;

but since the introduction of Steam boats, Ships have been dis-
used.

Towns. There are already more than 125 towns and villag-
es built on the Ohio. The city of Pittsburgh, at the head of it,
contains nearly 15000 inhabitants. Cincinnati, in Ohio, contains above 10,000. The other principal towns are: Louisville,
in Kentucky, at the falls, about 5000: Steubenville, in Ohio a-
bout 3000: Maysville or Limestone, in Kentucky, about 2000:
besides, Beavertown, in Pennsylvania: Wheeling, in Virginia:
Marietta, in Ohio, at the mouth of the Muskingum: Gallipo-
lis in Ohio: Portsmouth, Ohio, at the mouth of the Scioto: Au-
gusta, in Kentucky: Newport, K. at the mouth of Licking River: Owensborough, K. Hendersonville, K. Vevay, in Indiana:
Lawrenceburg, Ind. at the mouth of the great Miami: Madi.
son, Indiana: Jeffersonville and New-Albany, Indiana, both at
the falls: Evansville, Indiana: Shawneetown, in Illinois, &c.

Branches. The Ohio receives immediately about 400
streams, of which 20 are rivers above 100 miles long, 54 are
small rivers or large creeks, and more than 300 are brooks and
runs. Its largest branches empty into the lower parts of the River, such as the Tennessee, Cumberland, and Wabash. They
all flow in valleys similar to that of the Ohio and proportioned
to their size. Many of them, such as the Scioto, Miami, Tennessee, Wabash, &c. have plains, which indicate former lakes.
Most of them have rapids, ripples, bars, islands, &c. and offer
the same phenomena as the Ohio, particularly the periodical
rises and falls. I shall give some account of the 20 principal
streams, which fall into the Ohio, in the order in which they
join it.

PRINCIPAL BRANCHES OF THE OHIO.

1. Alleghany. It rises in Lycoming county, Pennsylvania,
near the 42d degree of latitude, on the northern parts of the
Alleghany mountains, and, after flowing through a small part
of the state of New-York, it returns into Pennsylvania, until
it joins the Monongahela at Pittsburgh and forms the Ohio.
General direction S. W. Length in a direct course 170 geographic miles, in the natural course 250, equal to 285 English
miles. It has five great branches, the Conemaugh, Conewa.

go, Tohas, &c. It is navigable throughout, and its stream is gentle and clear.

2. MONONGAHELA. Rises in the Alleghany mountains of Virginia, near latitude 38. Direct course N. and 150 miles, in the natural course 210 miles, or 245 English miles. It has three great branches, of which the Yohoghenv is the principal. Its breadth at Pittsburgh is 1350 feet, being wider and deeper than the Alleghany. It flows in a deep valley, is subject o sudden rises, and has a turbid but navigable stream.

3. MAHONING or BIG BEAVER. Rises near Lake Erie, in latitude 42, and runs south through Pennsylvania, emptying on the right side of the Ohio, of which it is one of the smallest branches, and is even sometimes called a Creek, although its direct course is 80 miles long, and the natural nearly 140, or about 163 English miles, being very crooked; but it is shallow, full of falls, and hardly navigable. It is formed by the junction of the Shenango and Neshanock.

4. MUSKINGUM. It flows through the state of Ohio, in a southerly direction, about 100 miles, but being very winding its natural course is 150 miles or about 175 English miles. It rises in a small lake of the Ohio ridge, which separates the bason of the Ohio from that of Lake Erie, near the 41st degree of latitude, and it joins the Ohio at Marietta. It is a large and navigable river, although it has a large rapid or fall at Zanesville and some other smaller rapids elsewhere. At the mouth it is 750 feet wide. It flows through a large valley, and receives four or five large branches, called Wills, Licking, Mohecan, &c.

5. LITTLE KENHAWAY. It rises in the Laurel hills, and flows through. Virginia in a N. W. course of 90 miles, or 140 in a natural course, equal to about 163 English miles. It empties at Parkenburg, is partly navigable and has several small branches.

6. HOCKHOCKING. Flows through Ohio. Direction, S. E. length seventy five miles, by the real course one hundred and twenty five, or about one hundred and forty English miles. It is a deep but narrow stream, navigable however as far as the two cascades. It had lakes formerly.

7. GREAT KENHAWAY. Rises in the Alleghany Mountains,

near latitude 36, in North Carolina, and flows through Virginia.' Course northerly, one hundreed and seventy five miles, real course very crooked, about two hundred and seventy miles or three hundred and fifteen English miles. It joins the Ohio a$_t$ Point Pleasant. It is a fine, navigable and broad river, with many branches.

8. BIG GUYANDOT. It rises in the Cumberland Mountains, and runs N. through Virginia, emptying itself at Guyandot It is navigable sixty miles; length seventy miles, real course one hundred miles, or about one hundred and twenty English miles.

9. SANDY RIVER. Rises also in the Cumberland Mountains near the 37th degree of latitude, and separates Virginia from Kentucky. It is a large but shallow river, with three branches. Common course north, ninety miles in length, natural course one hundred and twenty five miles, or one hundred and forty six English miles. It is also called Pottery river and Big Sandy.

10. SCIOTO. It flows through the state of Ohio, rising in a morass of the Ohio ridge or table land, near latitude 40 1-2. It empties near Portsmouth after a southerly course of one hundred and ten miles, real course about one hundred and ninety miles or two hundred and twelve English miles. It is navigable one hundred and thirty-miles, and is four hundred and fifty feet broad at the mouth. It has many bars and snags, but no falls. Its four principal branches are Whetstone river, Paint, Darby, and Walnut creeks. It had lakes formerly.

11. LITTLE MIAMI. Runs through Ohio in a S. S. W. direction of sixty miles, natural course one hundred miles or one hundred and fifteen English miles. It is not navigable. It joins the Ohio near Columbia and has several small branches. Near its head, it runs for a mile through a narrow chasm, with successive falls of two hundred feet.

12. LICKING RIVER. It flows through Kentucky in a N. W. course of one hundred and sixty miles, rising in the Cumberland Mountains, near latitude 37. It has two great branches, is hardly navigable, and winds very much. It empties between Newport and Covington, opposite Cincinnati. Real course about three hundred miles or nearly three hundred and fifty English miles.

13. GREAT MIAMI. It rises in the Ohio ridge, near latitude 40 1-2 and flows through Ohio in a S. S. W. direction, dividing that state from Indiana at its mouth, near Lawrenceburgh. Common course one hundred and ten miles, real course one hundred and eighty, or about two hundred and ten English miles. Its current is very rapid, and difficult to ascend. It has four principal branches, such as Mad river, Whitewater, &c. The mouth is six hundred feet wide, and its valley is very large. It was forerly called Rocky river.

14. KENTUCKY. This fine river gives its name to the state throughout which it flows, in a N. W. direction. It rises in the Cumberland Mountains, near the 37th degree of latitude, a high spot from which the Tennessee, Cumberland, Licking, &c. flow westward. Common course 180 miles, real course 340 and very winding, or about 400 english miles. It has 5 principal branches, Dick river, Black river, &c. It overflows in the spring and is then navigable even for Steam-Boats, &c. It has many rapids, but no real fall. Its valley is deep and often narrow; in the narrows, the limestone cliffs are 300 feet high, and very near each other, without any bottoms. It had formerly a few small lakes and hilly islands. It empties at Port William. Former name Cuttawa.

15. SALT RIVER. Flows in Kentucky, rises in the knobby hills, course N. W. 80 miles long, natural course winding about 140 miles, or 160 english miles. It is partly navigable and has many branches. It empties at Adamsville.

16. GREEN RIVER. It rises in Kentucky, in the knobby hills, which are spurs of the Cumberland Mountains, and flows West and N. W. into that state. Direct course 175 miles, usual course about 350 or more than 400 english miles. It has four large branches, such as Barren river, Rough and Panther creeks, &c. It has a gentle current and is navigable. Its valley is very wide in the lower part, and when it joins the Ohio, above Evansville, its stream is almost as large as the Ohio. It was formerly called Buffaloe river.

17. WABASH. It rises in Indiana, on the ridge dividing the basons of the Ohio and the Lakes, near latitude 41½, and below it forms the limits between Indiana and Illinois. Direction S.

S. W. Length 250 miles, real course 450 miles or nearly 525 English miles. It is a large and deep stream, navigable even in summer, as far as the falls. Its lower valley is wide and shal_ low, with many islands and bayous. It has five large branches, such as Little Wabash, White river, &c. This last is very considerable and extends its numerous and large branches throughout Indiana ; the longest is 350 miles long, one of them runs parallel with the Ohio. It empties above Shawneetown.

18. SALINE RIVER. It flows through Illinois in a S. E. direction, emptying below Shawneetown. Length 55 miles, rea_t course about 90, or 105 English miles; it is therefore the smal_ lest of the rivers emptying into the Ohio, although Big Blue river, Tradewater river, Little Muskingum, and Little Scioto, are still smaller and rather large creeks ; their course being less than 100 miles, I have not noticed them. The Saline river is partly navigable and has three principal branches.

19. CUMBERLAND. It rises in the Cumberland Mountains of Kentucky, and after watering Tennessee, returns into Kentucky, its course being W. and N. W. about 300 miles ; real course a- bout 500 miles or about 585 English miles. It is a fine naviga_ ble river, flowing in a broad valley, and with many small branch_ es, but no large ones. It has also been called the Shawanee.

20. TENNESSEE. The last and largest of the branches of the Ohio. It is formed by the union of the Holstein and Clinch riv_ ers in Tennessee, the former rising in Virginia near lat. 37, and the second in North Carolina, within the Alleghany Mountains near lat. 35. The whole course, if the Clinch river is deemed the main branch, will be three hundred and fifty miles, and the real course six hundred and fifty, equal to about seven hundred and sixty english miles. Duck river is another large branch of it, and there are three others besides. The direction is S. W. then west and next north, watering Tennessee, Alabama, Kentucky, &c. and emptying into the Ohio a few miles below the Cumberland, from which basin it is divided by a high ridge, and not far above the mouth of the Ohio. The Tennessee is a very large and fine navigable river, almost equal to the Ohio in size, but not in depth. Its valley is wide and has had many lakes, one of them was at the Muscle Shoals, which forms now a

small lake, full of rocky islands and rapids, and are a great impediment to navigation. It was formerly called the Cherokee river.

SMALLER BRANCHES.

The fifty four small rivers and large creeks, flowing into the Ohio are the following, of which thirty three empty on the right and twenty one on the left. They are all over thirty miles long in their natural course.

In PENNSYLVANIA, 3. Right bank, Little Beaver; and on the left bank Chartier's Creek, Raccoon Creek.

In OHIO, 17. Big Yellow creek, Warren creek, Indian Wheeling creek, Captina creek, Sunfish creek, Opossum creek, Little Muskingum river, Duck creek, Shade river, Kaygers creek, Campaign creek, Raccoon creek, Symmes' creek, Brush creek, Little Scioto river, Eagle creek, White Oak creek.

In VIRGINIA, 7. Short creek, Wheeling creek, Big Grave creek, Fishing creek, Stony creek, Big Sandy creek, Little Guyandot river.

In KENTUCKY, 12. Little Sandy river, Tygert creek, Kinniconick, Gunpowder creek, Bigbone creek, Harrod creek, Beargrass creek, Otter creek, Sinking creek, Blackford creek, Highland creek, Tradewater river.

In INDIANA, 12. Tanner's creek, Houghan creek, Loughery creek, Indian Kentucky, Silver creek, Buck creek, Corydon creek, Big Blue river, Little Blue river, Anderson river, Little Pigeon creek, Big Pigeon creek.

In ILLINOIS, 3. Lusk's creek, Bigbury creek, Cash river.

FISHES OF THE OHIO.

FIRST PART. THORACIC FISHES.

Having complete gills, with a gill cover, and a branchial membrane. Lower or ventral fins situated on the breast or thorax, under the pectoral or lateral fins.

1 GENUS. PERCH. PERCA. Perche.

Body elliptical, scaly; head without scales, mouth large, jaw with unequal acute teeth, gill cover with a serrate preopercule and a spiny opercule; two dorsal fins, the first with spiny rays, the second with soft rays. Vent posterior.

This genus was very badly defined by Linneus, Shaw, Bloch, and Mitchell; the above characters are now precise and apply to all the species that ought to . remain in it, answering to the genus of Lacepede and the subgenus of Cuvier, bearing the same name. All the species belonging to it are voracious and prey on smaller fishes. There are only few species in the Ohio, which afford an excellent food.

1st Species. SALMON PERCH. *Perca Salmonea*. Perche Saumonce.

Jaws nearly equal, one spine on the opercule and another at the base of the pectoral fins: body lengthened, breadth one ninth of the length, brownish above, with gilt shades, white beneath first dorsal fin with fourteen rays, second with twenty, tail fork ed, all the fins spotted; lateral line diagonal and slightly curved.

A fine fish, from one to three feet long; it is one of the best afforded by the Ohio, its flesh is esteemed a delicacy, being white, tender, and well flavoured, whence the name of *Salmon* was given to it, and its shape which is nearly cylindrical and slightly compressed, with the head and jaws somewhat similar to those of the Salmons, has induced many to consider it a real Salmon, although its fins and gill covers are quite different. It has received the vulgar names of *Salmon*, *White Salmon*, and *Ohio Salmon*. It is not a common fish, but is occasionally caught all over the Ohio and in the Kentucky, Licking, Wabash, and Miami rivers during the spring and summer; it appears at Pittsburgh sometimes as early as February, while it winters in deep waters. It feeds on Chubs, Minnows, Suckers, &c. It is not often caught with the hook, but easily taken with the gig and seine. It has the back and sides gilt by patches, the head variegated with small gilt spots above and quite white beneath. The eyes are large, prominent and brown, situated above the corners of the mouth and surrounded with a gilt brown iris. The two dorsal fins are widely apart, the first ray of the first dorsal fin is short, the second dorsal fin is slightly falcate, they are both yellow as well as the tail and with brown spots, the other fins are pale yellowish with only a few brown

C

dotts. The rays are, in the anal 12, wherein the first is short, and spiny, thoracic 6, the first hardly spiny, pectoral 14, candal 20. The whole fish is covered with very small scales, and the lateral line begins above the opercule: the second spine outside of the opercule is remarkable.

2d Species. GOLDEN-EYES PERCH. *Perca chrysopis.* Perche œuil-d'or.

Upper jaw longer, one spine on the opercule, body oblong, breadth one fourth of total length, silvery with five longitudinal brownish stripes on each side, head brown above: lateral line diagonal and straight; first dorsal fin with eight rays, the second has 14, whereof one is spiny, tail forked, roseate, tip brown; base scaly.

Vulgar names Rock fish, Rock bass, Rock perch, Gold eyes, Striped bass, &c. It is commonly mistaken for the Rock fish or Striped bass of the Atlantic Ocean, the *Perca-Mitchelli* of Dr. Mitchell, (Trans. of the philos. Society of New York, vol. 1. page 413, tab. 3. fig. 4.) to which it is certainly greatly similar; but it differs from it, by the single spine of the opercule, the shape of the lateral line, the less number of stripes, the scaly tail, &c. It is not very common in the Ohio, and is hardly ever seen at Pittsburgh, being more common in the lower parts of the river, where it frequents the falls, ripples, and rocky shores. Its usual size is about one foot. It is very good to eat. It bites at the hook. The mouth is large with very small teeth, the three pieces of the gill cover are slightly crenulate, the middle one or preopercule being however deeply serrate. The eyes are large black with a large golden iris. The lateral line begins at the corner of the opercule and does not follow the curve of the back, the stripes are parallel with it and only two of them reach the tail. The branchial membrane has six rays; the spine of the opercule is not terminal. The dorsal fins are rufous and quite separate, the two first rays of the first are shorter, the second is brown posteriorly and diagonally, its base is scaly and such is also the base of the anal fin, which has similar colours, and 15 rays, whereof three are spiny. Pectoral fins with 16 rays. Thoracic fins incarnate with six rays, whereof one is spiny.

It will appear that this fish differs so widely from the forego: ing, as to be hardly reducible to the same genus; but its great similarity with the *Perca Mitchelli* has compelled me to retain it in this genus, notwitstanding many peculiar characters. I shall however venture to propose a new subgenus or section in the genus *Perca* for this fish, to which the *P. Mitchelli*, may perhaps be found to belong. It may be called *Lepibema* and distinguished by the scaly bases of the caudal, anal, and second dorsal fins, this last with some spiny rays, and all the three parts of the gill cover more or less serrulate, besides the small teeth.

The *Perca Salmonea* may also form a peculiar subgenus, or section distinguished by the cylindrical shape of the body, long head and jaws, large teeth, and a second spine outside of the opercule over the base of the pectoral fins. It may be called *Stizostedion*, which means pungent throat. I could have made peculiar genera of each of them, under the proposed names; but as they otherwise agree with the reduced genus *Perca*, I have preferred delaying this innovation until more species are found possessing the same distinctions, in which case my two perches may then be called *Stizostedion salmoneum, and Lepibema chrysops.*

3d Species. BLACK DOTTED PERCH. *Perca nigropunctata* Perche a-points-noirs.

Upper jaw longer, body brown, covered all over with black dotts, breadth one sixth of the length, lateral line nearly straight the anal fins very long, tail truncate. I have not seen this species, I describe it from a drawing made by Mr. Audubon. I am therefore doubtful, whether it is a real perch, particularly since the drawing does not show the serratures and spines of the gill cover. It might be a *Sciena*, or a *Dipterodon*, yet the shape of the body and the distant dorsal fins, induce me to rank it with the *G. Perca* until better known; when it may even turn out to be a peculiar genus, which the flexuose opercule, long anal fin and vent in the middle of the body, seem to indicate, and should it be a real perch, it must form a peculiar subgenus, which may be called *Pomacampsis* in either case. The vulgar names of this fish are Black Perch, Widow's Perch, Dotted Bass, Black Bass, Batchelor's Perch, &c. It is found only in the lower parts of the Ohio, from the falls to the mouth, and it runs up the small creeks, but is rare every where. Its length is from six to twelve inches. The snout is rounded, the head sloping and small, the preopercule rounded, the opercule flexuose or nearly lobate; the eyes are black and beyond the mouth. The back is almost black, the two dorsal fins are dotted like the body, the first has about twelve spiny rays, and the second about eight soft rays, this last is very near the tail. The anal fin has about twenty rays and begin just below the vent and the end of the first dorsal fin. Vent in the middle of the body, almost nearer the head.

2 — II Genus. Bubbler. Amblodon. Amblodon.

Body elliptical, compressed, scaly; head and gill covers scaly, jaws with small fily teeth, throat with a triangular bone beneath, covered with large round hollow and obtuse teeth. Gill cover with two pieces, preopercule slightly denticulate at the base, opercule without teeth nor spines: branchial membrane with six rays. Two dorsal fins contigous, the first spiny, the second partly so, scaly along the base. Vent posterior. - - _

This genus was called by me *Aplodinotus* G. 8, of my Memoir on 70 New Genera of American animals, in the journal of Natural History of Paris, having been led into error, in supposing that the remarkable teeth of its throat belonged to the Buffalo fish, as will be seen below. The name means obtuse teeth. It differs from the G. *Sciena* by the scaly head, opercule and base of second dorsal fin, besides the singular teeth. Only one species is known as yet.

4th Species. Grunting Bubbler. *Amblodon grunniens* Amblodon grognant.

Synonymy. *Sciena grunniens* Raf. Catal, fishes of Ohio; *Aplodinotus grunniens.* Raf. Mem. on 70 N. G. Animals, G. 8.

Entirely silvery, upper lip longer, lateral line curved upwards at the base, bent in the middle; and straight posteriorly, tail lunulate, first dorsal fin with nine rays, the first very short, the second with 35 rays, the first spiny and short.

The vulgar names of this fish are White-perch, White-pearch, Buffaloe-perch, grunting-perch, bubbling-fish, bubbler, and muscle-eater. It is one of the largest and best found in the Ohio, reaching sometimes to the length of three feet and the weight of thirty pounds, and affording a delicate food. It is also one of the most common, being found all over the Ohio, and even the Monongahela and Alleghany, as also in the Mississippi, Tennessee, Cumberland, Kentucky, Wabash, Miami, &c. and all the large tributary streams: where it is permanent, since it is found at all seasons except in winter. In Pittsburgh it appears again in February. It feeds on many species of fishes, Suckers,Catfishes, Sunfishes,&c. but principally on the muscles, or various species of the bivalve genus *Unio*, so common in the Ohio, whose thick shells it is enabled to crush by means of its large throat teeth. The structure of those teeth is very singular and peculiar, they are placed like paving stones on the flat bone of the lower throat, in great numbers and of different sizes; the largest, which are as big as a man's nails, are always in the centre; they are inverted in faint alveoles, but not at all connected with the bone; their shape is circular and flattened, the inside always hollow, with a round hole beneath: in the young fishes they are rather convex above and evidently radiated and mamillar; while in the old fishes they become smooth, truncate, and shining white. These teeth and their bone are common in many museums, where they are erroneously called teeth of the Buffalo-fish or of a Cat-fish. I was deceived so far by this mistake and by the repeated assertions of several persons, as to ascribe those teeth to the Buffalo-fish, which I have since found to be a real *Catostomus*; this error I now correct with pleasure.

A remarkable peculiarity of this fish consists in the strange grunting noise, which it produces, and from which I have derived its specific name. It is intermediate between the dumb grunt of a hog and the single croaking noise of the bull frog: that grunt is only repeated at intervals and not in quick succession. Every navigator of the Ohio is well acquainted with it, as they often come under the boats to enjoy their shade in summer and frequently make their noise. Another peculiarity of this fish, is the habit which it has of producing large bubbles in quick succession, while digging through the mud or sand o the river, in search of the Muscles or Unios.

It has a small-head, sloping and compressed all the way from the snout to the dorsal fins and entirely scaly; thick, hard, and extensible lips, and 2 nostrils on each side, the posterior larger oblong & obliqual: the opercule is rounded with gilt shades; those shades extend to the sides of the body, while the back is slightly dark or brownish, and the upper part of the head olivaceous. The iris is gilt brown and the fins are brownish, except the thoracic and pectoral, which are reddish; these last have 18 rays, while the thoracic have ~~some, whereof the~~ first is spiny and the second ~~wanting. Tail with~~ twenty rays. ~~And few~~ narrow elongate, ~~tow tinged~~ with reddish, and with nine rays, whereof ~~the first is~~ spiny, very small and flat, the second is also spiny, but very thick, large and triangular, the third ray is the longest and the last is mucronate. The first dorsal fin is triangular and broader than the second, which is very long and rounded behind.

This fish is ~~either~~ taken in the seine or with the hook and line; ~~it bites easily~~, and affords fine sport to the fishermen. It spawns in the spring, and lays a great quantity ~~of eggs.~~ —

III Genus. CALLIURUS. PAINTED TAIL. Calliure.

Body elongate, compressed, scaly; fore part of the head without scales, neck and gill-covers scaly: mouth large with strong teeth in both jaws, and without lips. Gill cover **double**, preopercule divided downwards into three curved and carinated sutures, without serrature: opercule with an acute and membranaceous appendage, before which stands a flat spine. One dorsal fin, spiny anteriorly, depressed in the middle. Anal fin with spiny rays, thoracic with none, and only five soft rays. Vent nearly medial.

The generic name means fine tail. It differs principally from the genus *Holocentrus*, by the head, scaly gill cover and singular propercule: Genus 12 of my 70 New Genera of American Animals.

5th Species. DOTTED PAINTED TAIL. *Calliurus Punctulatuse.* Calliure pointille.

Lower jaw longer: body olivaceous crowded with blackish dotts: head brownish, flattened above: lateral line hardly curved upwards at the base: tail unequally bilobed, lower lobe larger, base yellow, middle blackish, tip white: dorsal fin yellow with 24 rays, of which 10 are spiny.

An uncommon fish from four to twelve inches long. I observed it at the falls; rare in the Ohio, more common in some small streams. Vulgar names, Painted-tail or Bride-perch. Tail with two lobes, slightly unequal, base flexuose. Belly and lower fins pale, anal fin with 13 rays, the three anterior spiny and shorter, behind rounded and far from the tail, although nearer than the dorsal fin. Thoracic fin with five rays, none of which appear spiny, and no appendage. Pectoral fins short, trapezoidal, with 15 rays. Branchial rays concealed.

IV. Genus. SUNFISH. ICTHELIS. Icthele.

Body elliptical or oval very compressed, scaly. Mouth small, with small teeth and thin lips. Gill cover double, scaly, without serrature or spines. One dorsal fin, broader behind with anterior spiny rays, as well as the anal and thoracic fins, these without appendages. Vent hardly posterior. Lateral line following the curve of the back.

Synonomy *Lepomis*. Prod. 70 New Genera, 13 Genus.

An extensive genus, which contains perhaps 20 species, most of which were blended with the *Labrus auritus* and *Labrus virginicus* of Linneus. They differ from the genus *Labrus* or rather *Sparus*, by the scaly opercule and the thoracic fins without appendage. I have already detected six species in the western waters; but there are more. I divide them into two subgenera. Meaning Sun-fish - All good to eat, and easily taken with the hook; they feed on worms and young fishes. They are permanent.

1st Subgenus. TELIPOMIS.

Opercule without appendage; but spotted—Meaning spotted gills.

6th Species. GILDED SUNFISH. *Icthelis macrochira*. Itchele macrochire.

Body oval,oblong, gilt, crowded with small brown dotts; head small, scaly, opercule flexuose, spot narrow marginal and black, jaws equal: tail forked: pectoral fins long and narrow, reaching the anal fin, which has 13 rays, whereof three are spiny.

A pretty species from three to four inches long. In the Ohio, Green river, Wabash, &c Names, Sun-fish, Gold-fish, &c. Head rather acute, not scaly before the eyes. Iris gilt brown.

Dorsal fin with 22 long rays, whereof 11 are spiny, a depression between the two sorts of rays. Anal fin broad and rounded. Tail 20 rays. Thoracic one and five. Pectoral 15. Diameter of the body nearly one fourth of total length.

7th Species. **BLUE SUNFISH.** *Icthelis cyanella.* Icthele bleuatre.

Body elliptic, elongate, diameter one fifth, olivaceous gilt, crowded with irregular ~~blue dots, brownish~~ above: head elongate, lo~~wer jaw longer, stroke with~~ blue flexuose lines; spot oblong ~~blackish, nearly~~ marginal: tail rounded, notched: anal fin ~~very broad~~ with 12 rays, whereof three are short spiny: pectoral fins very short.

A small species hardly three inches, called Blue-fish or Sunfish. I found it on the Ohio at the falls. Appearing entirely blue at a distance. Head brown above. Iris gilt. Opercule curved. Tail olive blue, with 24 rays. Dorsal fin brownish with 20 rays, ~~whereof 10 are spiny, hardly any~~ middle depression. Pectoral small trapezoidal, 12 rays. Thoracic one and five.

8th Species. BLACKEYE SUNFISH. *Icthelis melanops* Icthele œuil-noir.

Body oblong, diameter one ~~fourth,~~ olivaceous covered with blue dotts, neck brown above, head large, ~~mouth rather large,~~ upper jaw longer; opercule with blue curved and longitudinal lines beneath: spot rounded black at its base: fins olivaceous, tail bilobed: anal fin with ~~three and~~ nine rays: pectoral fins large oboval.

Length from two to six inches: common in the tributary streams of the Ohio, the Kentucky, Licking, Miami, &c. and even in small creeks. Vulgar names, Blue-fish, Black-eyes, Sun-fish, Blue-bass, &c. It has black eyes like all the other species; but the iris is black also, with a silvery hue or ring. Dorsal fin with ten and ten rays, the spiny ones very short. Caudal 20. Pectoral 16. Thoracic one and five, as usual; but the spiny ray is very short, as are also those of the anal fin.

2d Subgenus. POMOTIS.

Opercule with a membranaceous appendage, often like an auricule and spotted. Meaning eared gills.

9th species. REDEYE SUNFISH. *Icthelis Enythrops.* Itchele œuil rouge.

Body oval elliptic, (diameter one third,) blackish above, olivaceous on the sides, whitish beneath: head small, lower jaw longer, preoperculefl exuose, opercule with a short, angular and acute appendage, a faint and small brown spot above it: tail bilobed cillate, base black, middle olivaceous, tip whitish, upper lobe rather larger: anal fin with six and ten rays: pectorals trapezodial large, not reaching the vent.

Vulgar names Red-eyes, and Sunfish. Observed in Licking river, said to be common in many other streams. Length 3 to 8 inches. All the fins olivaceous. Eyes black, iris large and red. Dorsal 11 and 10 rays, spiny short, as well as the 6 of the anal fin. Caudal 17. Pectorals 12. Thoracics 1 and 5, the spiny ray long.

10th Species. EARED SUNFISH. *Icthelis aurita.* Icthele oreillee.

Body oval elliptic (diameter one third) olivaceous with blue and rufous dots: head small, jaws equal, opercule flexuose, appendage black, broad and truncate, some blue flexuose lines on the sides of the head: tail brownish lunulate; back brownish: anal fin 3 and 9: pectorals not reaching the vent. Thoracic mucronate.

Length from 3 to 12 inches: common in the rivers, creeks, and ponds of Kentucky. Vulgar name Sunfish. Iris brown. Dorsal fin brownish, 10 and 10, spiny rays shorter. Thoracic fins very long, spiny ray rather shorter, first soft ray mucronate. Pectorals nearly rhomboidal, with 14 rays. Tail 16 rays.

11th Species. BIG-EAR SUNFISH. *Icthelis megalotis.* Icthele megalote.

Body oval rounded, (diameter two fifths,) chesnut colour with blue dots, belly red: head large, lower jaw longer, opercule with blue flexuose lines, appendage black, very large elliptic, end rounded: tail black, slightly forked: pectoral large reaching the vent: anal fin 3 and 9: thoracics long and mucronate. Black tail.

A fine species, called Red-belly, Black-ears, Black-tail Sunfish, &c. It lives in the Kentucky, Licking, and Sandy rivers, &c. Length from 4 to 8 inches. Head very sloping. Iris sil very brown. Belly of a bright copper red colour. All the fins black except the pectorals which are olivaceous, trapezoida acute and large. The dorsal has 20 rays, whereof 9 short ones are spiny. Body very short, hardly as long as broad, if the head and tail are deducted. Thoracics like those of the forego ing species

V Genus. River Bass. Lepomis. Lepome.

This genus differs from ~~Holocentrus~~ by having the opercule scaly, from *Calliurus* by the opercule only being such, while the preopercule is simple and united above with a square suture over the head, besides the thoracic fins with 6 rays. Perhaps the *Calliurus* ought only to be a subgenus of this. From the G. *Icthelis* it differs by the large mouth and spines on the oper cule.

The name means scaly gills. The species are numerous throughout the United States. They are permanent; but ramblers in the Ohio and tributary streams. They are fishes of prey, and easily caught with the hook. I shall divide them into two subgenera. I had wrongly blended this genus and the Icthelis under the name *Lepomis* 13. G. of my Prodr. N. G.

- 1st Subgenus. Aplites.

Only one flat spine on the opercule, decurrent in a small medial opercule: first ray of the thoracic fins soft or hardly spiny. Meaning, single weapon.

12th Species. Pale River-bass. *Lepomis pallida*. Lepome pale.—

Olivaceous above, white beneath, a brown spot at the base of he lateral line, an obtuse appendage on the opercule, spine behind it: 3 faint oblique streaks on the gill covers; lower jaw longer: tail forked, pale yellow, tip brown.

Not uncommon in the Ohio, Miami, Hockhocking, &c. Vulgar name Yellow Bass, common Bass, &c. Length from 4 to 12 inches. Shape elliptic, diameter one fourth of the total length. Fins olivaceous, without streaks, dorsal depressed or interrupted in the middle, 9 spiny rays to the fore part, the medial longer, 1 spiny ray and 14 soft rays to the hind part. Anal fin rounded 13 rays, whereof 2 are spiny and short. Pectorals rounded with 14 rays. Tail with 18. Thoracics with 6. Eyes large, black, iris brown with a gold ring. Lateral line following the back, straight near the tail.

13th Species. STREAKED-CHEEKS RIVER-BASS. *Lepomia trifasciata.* Lepome trifasciee.

Whitish, crowded with unequal and irregular specks, of a gilt olive colour, none on the belly: gill covers with 3 large oblique streaks of the same colour: opercule without appendage, spine acute, a faint brown spot below the lateral line: lower jaw longer: dorsal fin streaked behind: tail forked, yellow at the base, brown in the middle, tip pale.

Found in the Ohio and many other streams, reaches over a foot in length sometimes: vulgar names Yellow bass, Gold bass, Yellow perch, Streaked-head, &c. Fins olivaceous: dorsal hardly depressed in the middle with 34 rays, whereof 10 are spiny, hind part with 3 brownish and longitudinal streaks. Anal fin rounded with 13 rays, 3 of which are spiny, 2 short and a long one. Pectoral fins nearly triangular and acute, 16 rays. Thoracics 6. Tail 2, very broad, forks divaricate nearly lunulate. Eyes small black, iris brown. Lateral line following the back. Diameter less than one .ourth of the length.

14th Species. BROWN RIVER-BASS. *Lepomia flexuolaris.* Lepome flexueux.

Olivaceous brown above, sides with some transversal and flexuose olive lines, belly white: lateral line nearly straight flexuose: spine broad acute, behind the base of the opercule, no appendage nor spot, preopercule forked downwards: upper jaw slightly longer: tail bilobed, base olive, middle brown, tip white.

A fine species, reaching the length of two feet, and affording an excellent food. Common all over the Ohio and tributary streams. Vulgar names Black Bass, Brown Bass, Black Pearch, &c. Fins olivaceous, dorsal with 23 rays, whereof 9 are spiny and rather shorter: anal with 12 rays, whereof 2 are spiny: pectorals trapesoidal, 16 rays. Branchial rays uncovered. Iris brown. This fish might perhaps form another subgenus, by the large mouth, head without upper sutures, spine hardly decurrent, nearly equal jaws, gill covers, lateral line, &c. Its tail and preopercule are somewhat like *Calliurus.* It might be called *Nemocampsis,* meaning flexuose line. Diameter one fourth of the length.

2d Subgenus.   DIOPLITES.

Opercule with two spines above. First ray of the thoracic fins spiny. Lateral line curved as the back. Meaning two weapons.

15th Species. TROUT RIVER-BASS. *Lepomis Salmonea*. Lepome saumone.

Olivaceous brown above, sides pale with some round yellowish spots, beneath white; preopercule simple, head without sutures, lower jaw slightly longer, spines flat, short, acute, and decurrent above and beneath, opercule acute beneath the spines whitish, tip blackish: vent posterior.

Length from 6 to 24 inches. Vulgar names White Trout, Brown Trout, Trout Pearch, Trout Bass, Brown Bass, Black Bass, Black Pearch, &c. Common in the Kentucky, Ohio, Green, and Licking rivers, &c. It offers a delicate white flesh, similar to the *Perca Salmonea*. It is a voracious fish, with many rows of sharp teeth on the jaws and in the throat. It bites easily at the hook, and eats suckers, minnows, and chubs. Diameter one fifth of the length. Fins olivaceous brown; dorsal with 25 rays, whereof 10 are spiny, slightly depressed between them: anal rounded small, 8 and 11 rays. Pectoral acute trapesoidal 18 rays. Thoracic rays, spiny ray half the length. Tail with 24 rays. Iris silvery.

16th Species. SPOTTED RIVER-BASS. *Lepomis notata*. Lepome tache.

This species differs merely from the foregoing, by having a black spot on the margin of the opercule, two diagonal brown stripes on each side of the head below the eyes, and all the fins yellow, except the tail which is black at the end, with a narrow white tip. It is also smaller, from 3 to 8 inches long. It bears the same vulgar names and is found along with it, of which some fishermen deem that it is the young. But I have seen so many false assertions of the kind elsewhere, that I am inclined to doubt this fact, as it would be very strange that the gradual changes should be so great. Yet this ought to be enquired into, since many vulgar opinions are often found to be correct.

17th Species. SUNFISH RIVER-BASS. *Lepomis ictheloides*. Lepome ictheloide.

Silvery, olivaceous above, some faint blackish spots on the sides: lower jaw hardly longer, head with sutures, two flat, broad and obtuse spines above the opercule, decurrent with the sutures. Vent medial. Tail lunulate. Diameter one fourth of the length.

A very distinct species from the two foregoing. It might almost form a peculiar subgenus, by the medial vent, and obtuse spines situated above the lateral line and opercule. It might be called *Amblofilites* or obtuse weapons. It is found in the Kentucky and tributary streams. Vulgar names White Bass, or Sunfish Bass. Length from 4 to 8 inches. It is also a fish of prey and has many rows of sharp teeth. Its flesh is like that of the Sunfishes. Lateral line following the curve of the back. Iris silvery. Body with gilt shades; dorsal with 21 rays, 11 spiny, no depression. Anal 15, whereof 5 are spiny and gradually shorter. Thoracics 1 and 5. Pectoral broad 12 rays. Tail 16. Branchial rays 5. A faint and narrow marginal black spot on the opercule beneath the spines.

VI Genus POMOXIS. POMOXIS. Pomoxe.

Body elliptic, compressed, scaly. Vent anterior. Head scaleless, jaws plaited extensible, roughened by very minute teeth. Gill cover smooth, scaleless, propercule forked beneath, operculo membranaceous and acute posteriorly. Thoracic fins without appendage, but a spiny ray. One dorsal fin opposite to the anal, both with many spiny rays.

A very remarkable genus by the anterior vent, equal anal and dorsal fin, by which it differs from the genus *Sparus*, besides the want of appendage, &c. The name means acute opercule.

18th species. GOLD-RING POMOXIS. *Pomoxis annularis.* Pomoxe annulaire. -

Synonymy. *Pomoxis annularis.* Journal of the Acad. of Nat. Science of Philadelphia, vol. 1, p. 417, tab. 17, fig. 1.

Silvery, back olivaceous, with some geminate brown transversal lines; a golden ring at the base of the tail; lateral line straight: dorsal and anal fins with six spiny rays, a marginal black spot behind both fins: tail forked: lower jaw longer.

Vulgar names Gold-ring and Silver-perch. Found in August at the falls, probably permanent, Length from three to six inches. Good to eat. Eyes black, iris silvery. Diameter three tenths of the length. Head gilt above. Pectoral fins reaching the vent Scales deciduous and a little ciliated. End of the tail blackish. Spiny rays of the anal and dorsal fins gradually longer, but shorter than the soft rays, which are also gradually decreasing; the dorsal has only 14, while the anal has 16 such rays. Caudal 28. Thoracic one and five.

VII Genus. RED-EYE.- APHOCENTRUS. *Aplocentre.*

Body elliptic, compressed. Head small, jaws with lips and teeth, opercule smooth and flexuose. Vent medial. One longitudinal dorsal fin with only one spine.

A singular genus, intermediate between *Labrus, Cynedus,* and *Coryphena;* but belonging to the family of Labrides. The name means single spine. I describe it from a drawing made by Mr. Audubon. It is also the 11th genus of my Prodromus.

19th Species. OHIO RED-EYE. *Aplocentrus calliops.* Aplocentre belæuil. —

Pale greenish above, with some flexuose transversal black lines, yellowish beneath the lateral line, and with some small black lines, whitish and unspotted beneath: iris red: forehead flexuose convex: upper jaw hardly longer: dorsal spine longer: tail flabelliform: lateral line straight.

A beautiful fish from eight to twelve inches long. It lives in the lower parts of the Ohio, in Green river, &c. Vulgar names Red-eyes, Bride pearch, Batchelor's pearch, Green bass, &c. Breadth about one fourth of the length. Dorsal fin beginning behind the head with a long spiny ray and ending close to the tail, variegated with flexuose black lines: broad at the base, depressed near the tail, and suddenly broad again at the end. Anal fin small. Thoracic fin triangular. Lateral line rather broad. Iris large and red. Tail unspotted, and with rounded tip or fan-shaped.

VIII Genus. BARBOT. POGOSTOMA. Barbotte.

Body oval, compressed. Head small, jaws equal, without teeth, but with lips and six barbs, two to each lip and two to the lower jaw: opercule smooth, rounded. Two distant dorsal fins.

A fine genus next to *Dipterodon* and *Cheilodipterus;* it bolongs to the family of Labrides, and is distioguished from all the other genera by its barbs. The real name means bearded mouth It was the 10th genus of my prod. of 70 new genera.

20th Species. WHITE-EYES BARBOT. *Pogostoma leucops.* Barbotte œuilblanc.

Brown, with five black curved streaks, two on each side and one on the back, lateral line curved joining the lower streak: whitish beneath; a row of transversal lunulate, geminate and black lines, between the two lateral streaks, six similar ones on the gill cover: a large white and round patch surrounding the eyes: tail forked: vent posterior.

A beautiful fish: shape of sunfish: length sometimes twelve inches and weight one pound. It is found in the lower part of the Ohio and in the Mississippi; but is a rare fish. It has great many vulgar names, such as White-eyes, Spectacles-fish, Streaked Sunfish, Black Sunfish, Barbot, Bear ed Sunfish, &c. and the French settlers call it Barbotte, Poisso : l inette, and Œuil blanc. It does not bite the hook, and is only taken with the seine. The row of lunulated lateral lines have the convexity towards the head and extend through the tail. The two dorsal fins are short and trapezoidal, anal fins very small. Pectoral long. Thoracic under their hind part. Convexity of the three pairs of lines on the opercule, looking upwards. Eyes small and black, iris narrow and yellow, the white patch appears as a second iris. Chin and forehead between the eyes depressed, which form a kind of rounded snout, mouth small, jaws c qual. I describe it from a drawing of Mr. Audubon.

IX Genus. HOGFISH. ETHEOSTOMA. Etheostome.

Body nearly cylindrical and scaly. Mouth variable with small teeth. Gill cover double or triple unserrate, with a spine on the opercule and without scales: six branchial rays. Thoracic fins with six rays, one of which is spiny; no appendage. One dorsal fin more or less divided in two parts, the anterior one with entirely spiny rays. Vent medial or rather anterior.

A singular new genus, of which I have already detected five species, so different from each other that they might form as many subgenera. Yet they agree in the above characters, and differ from the genus *Sciena* by the shape of the body and mouth, and the divided dorsal fin. The name means different mouths. I divide it into two subgenera. They are all very small fishes, hardly noticed, and only employed for bait; yet they are good to eat, fried, and may often be taken with baskets at the falls and mill races. They feed on worms and spawn.

1st Subgenus. APLESION.

Dorsal fin single, split in the middle. Meaning nearly simple

21st Species. BASS HOGFISH. *Etheostoma calliura.* Etheostome calliure.

Body slightly fusiform and compressed, silvery, olivaceous above, some flexuose transversal brownish lines on the sides: lower jaw longer, preopercule double, opercule with an angular appendage and an obtuse spine behind it: scales smooth, lateral line flexuose tail forked, tri-coloured, and with a brown spot at the base.

The largest species of the genus from three to nine inches long. It has some similarity with the *Lepomis flexuolaris,* and some other River bass, wherefore it is called Minny-bass; Little bass, Hog-bass, &c. common in the Ohio, Salt river, &c. It has sharp teeth. The head is large, rugose above: iris large-gilt brown: branchial rays uncovered. Diameter one seventh of the length. Lateral line curved upwards at its base. Fins olivaceous. Dorsal with 9 and 14 rays, beginning behind the pectorals and ending far from the tail, like the anal, which has 12 rays, whereof one is spiny. Pectoral fins short trapezoidal 16 rays. Tail 24, fine, base with a yellow curved ring, followed by a forked band of a pale violaceous colour, tip hyalin. Mouth straight.

22d Species. FANTAIL HOGFISH. *Etheostoma flabellata.* Etheostome eventail.

Body olivaceous brown, with transverse unequal brown streaks, a black spot at the lower base of the lateral line which is straight; scales ciliated: mouth puckered obliqual, jaws nearly equal, cheeks swelled, preopercule simple, opercule curved, spine acute: pectoral fins rounded. Tail oboval flabelliform.

A small fish only two or three inches long, common at the falls of Ohio. Vulgar names Fan-tail, Black bass, Pucker, &c.

Head small, with swelled and dotted cheeks: iris brown with an internal gold ring; branchial rays concealed.. ▊▊▊▊ small roughened. Dorsal fin beginning above the pectorals ▊▊ end ing beyond the anal, with 8 short spiny rays and 12 soft ones olivaceous, with a longitudinal brown stripe. Vent anterior-anal fin very far from the tail, convex pale, rays 1 and 8. Pec, toral fins 15. Caudal 20, olivaceous with many small transver-sal ▊▊▊▊ lines. Diameter less than one seventh of the length.

23d Species. BLACK HOGFISH. *Etheostoma nigra.* Etheos-tome noire.

Entirely black, pale beneath; scales smooth, lateral line streight, mouth rather beneath, forehead rounded, upper jaw longer; preopercule rounded, ▊▊▊▊ anteri-or: tail entire nearly truncate. From one to two inches long. Observed in Green river. Vulgar name Black minny. Iris black, silvery, and small. Di-ameter one seventh of the length, without spots. Head small. Pectoral fins oboval. Tail 20. Anal fin 2 and 8. Dorsal 10 and 12.

2d Subgenus. DIPLESION.

Dorsal fin nearly double, divided into two joining parts. Meaning nearly double.

24th Species. BLUNT NOSE HOGFISH. *Etheostoma ▊▊▊oi-des.* Etheostome ▊▊▊▊

Body elongate, breadth one eighth of the length, olivaceous almost diaplanous, some brown spots on the back, and some brown geminate transversal lines across the lateral line which is straight, but raised at the base. Head small, snout rounded, mouth small beneath, lower jaw shorter; opercule angular, ▊▊▊▊ciliated, pectoral fins elongate, tail also, and ▊▊▊▊

A strange species ▊▊▊▊ of many Blennies. ▊▊▊▊ Seen in the Ohio, Wabash, Muskingum, &c. Colour pale, sometimes fulvous, whitish beneath. Cheeks swelled and smooth, preopercule simple arched, opercule quite angular: iris large and blackish: scales roughened by the ciliation. Dor-sal fin 13 and 13, beginning above the middle of the pectorals and ending with the anal, one faint longitudinal brown stripe on

E

&c. Tail 20 rays, with many small transversal lines. Vent medial. Anal fin 2 and 8. Pectoral fins 16, oblong acute.

25th Species. COMMON HOGFISH. *Etheostoma caprodes.*
Etheostome capros.

Body quite cylindrical, whitish, with about twenty transverse bands, alternately shorter. Head elongate obtuse, upper jaw longer, rounded; opercule angular, spine acute: lateral line quite straight: diameter one eighth of the length: tail forked, olivaceous ⬛⬛⬛⬛⬛⬛⬛ ⬛⬛⬛ with a black dot. Vent anterior ⬛⬛⬛⬛⬛⬛ ⬛⬛⬛⬛⬛⬛⬛ common species, found in the Ohio, Cumberland, Wabash, Tennessee, Green River, Kentucky, Licking, Miami, &c.; called almost every where Hog-fish. Length from two to six inches. Scales rough upwards, hardly ciliate. Mouth beneath, small; upper jaw protruding like a hog's snout, the nostrils being on it. Eyes above the eyes, jutting, black, iris silvery. Sides of the head silvery, above fulvous; preopercule simple arched. Branchial rays half visible. Fins hyalinous: dorsal 15 and 12, ending before the anal, which is very distant from the tail, rays 2 and 10. Pectoral fins trapezoidal 16. Tail 24.

SECOND PART. ABDOMINAL FISHES.

Having complete gills, with a gill cover and a branchial membrane. Lower or ventral fins situated on the belly or abdomen, behind the pectoral or lateral fins.

X. Genus. GOLDSHAD. POMOLOBUS. Pomolobe.
Body nearly cylindrical, elongate, scaly. Vent posterior. Abdomen carinated and serrated from the head to the vent; but without plaits or broad scales. Head scaleless, opercule lobed, with a rounded shield above. Jaws without teeth, upper truncate extensible, lower horizontal and fixed. Abdominal fins with nine rays and without lateral appendage: dorsal fin opposite.

Out of eight species of fishes, similar to the Herrings and Shads, which have already been observed in the Ohio; after an attentive study, I have ascertained that not a single one is a real Herring, nor belongs to the genus *Clupea*, and I have been compelled to form four new genera with them; which afford ⬛⬛⬛⬛ characters. The present one differs from the ⬛⬛⬛ genus *Clupea* by the lobed and shielded opercule, the curious mouth, the bodily shape, and the want of lateral appendage. It belongs of course, with the four following, to the family of Clupides. The name means lobed opercule.

26th Species. OHIO GOLDSHAD. *Pomolobus. chrysochloris.* Pomolobe doré.

Greenish-gold above, silvery beneath; lateral line straight: diameter two ninths of the length: dorsal and anal fin trapezoidal and with 18 rays: tail brown and forked.

A fine fish from twelve to eighteen inches long. Flesh esteemed; white and with less bones than the shad. It is taken with the seine and harpoon, as it seldom bites at the hook; it preys however on some small fishes. It seldom goes as far as Pittsburgh, and does not run up the creeks. In the falls it appears in March and April, and disappears in September. Its vulgar names are Ohio Shad, Gold Shad, Green Herring, &c.

It has the back convex, blue under the scales. Sides, belly, and throat with purple and violet shades. Top of the head and neck clouded with brown. Several sutures on the sides of the head. Upper lip truncate, flexuose, and even retuse; the lower obtuse and brown at the end. Eyes black; iris silvery and gilt. Opercule nearly trilobe, the upper lobe covered by a large oboval and radiated shield. Scales large deciduous, lateral line concealed by them. Dorsal fin olivaceous, in the middle of the back, first and second ray shorter and simple, the third long, the others gradually shorter. Anal fin consimilar but whitish. Pectoral and abdominal fins trapezoidal, the lowest ray simple and the longest: pectoral 15 rays. Tail equal 32 rays, brown, tip darker, some decurrent on each side; end of the body truncate.

3d Genus—GIZZARD. DOROSOMA. Dorosome.

Body lanceolate, compressed, scaly. Vent medial. Abdomen carinated, serrated, and with broad transversal scales, as far as the abdominal fins. Head scaleless, gill cover triple, opercule simple: mouth diagonal without teeth, lower jaw shorter. Abdominal fins with nine rays and no appendage: dorsal opposite.

It differs from *Clupea* and *Pomolobus*, by the medial vent, lanceolate body, gill covers, &c. The name means lanceolate body.

27th Species. SPOTTED GIZZARD. - *Dorosoma notata.* Dor- osome tachee.

Entirely silvery, a large brown and round spot above the base of the lateral line, which is straight: two oblong spots of an em- erald colour above the head: dorsal fin trapezoidal with 15 rays, anal longitudinal with 40. Tail unequally forked, lower lobe longer.

A small species, ... or ten inches. Diameter .... gradually .... I found it below the falls of the Ohio .... also in the spring and disappears in the fall: Vulgar nomes Gizzard, Hickory Shad, White Shad, &c. It does not bite at the hook. Back faintly bluish. Mouth large, upper jaw obliqual straight and longer, both fixed: tongue long and smooth. Eyes large, bluish, with a black centre: iris sil- very. Scales small. Pectoral 12 rays, abdominals immediately behind them ....

.... GOLD HERRING. NOTEMIGONUS. .... Body fusiform, compressed, scaly. Vent posterior. Abdo- men obtusely carinated, not serrate. Back similar before the dorsal fin. Head scaleless, mouth .... teeth, lower jaw longer: gill cover double, opercule simple .... fins with nine rays and no lateral appendage. Dorsal fin behind them above the vent.

This genus differs from *Clupea* by the carinated back and belly, without serratures, and the posterior do .... means back half-angular. 14th G. of my .... of an- imals. ....

.... *Notemigonus aura- .... dore.* ....

Back gilt olivaceous, remainder gilt silvery: fins yellow; lat- eral line following the curve of the belly: dorsal with 8 rays, anal with 12: tail equally forked. ....

Length from four to eight inches, diameter one fifth of the total length. Iris gilt. Tongue short, toothless, Scales large radiating with nerves. Head convex above and small. Dorsal fin broad trapezoidal, the first ray longer. Anal broad also, but .... Pectoral small with 16 rays. Tail 24. N.........

common in the Ohio, Kentucky, Miami, &c. The vulgar names are Gold Herring and Yellow Herring. It appears in the fall. It does not bite at the hook. Flesh pretty good.

XIII Genus. FALSE HERRING. HYODON. Hyodon.

Body lanceolate or oblong, compressed, scaly. Vent posterior. Abdomen slightly and obtusely carinated between the abdominal fins and the vent. Head scaleless: mouth toothed all over, strongly on the tongue, which is formed by the hyodal bone; lower jaw narrow and commonly longer. Gill cover with a preopercule. Abdominal fin with seven rays and a lateral appendage. Dorsal fin behind them above the base of the anal fin.

Hyodon. Lesueur in Journalof the Academy of Natural Sciences of Philadelphia, vol. 1 page 366.

Glossodon. Rafinesque in American Monthly Mag. 1818.

Amphiodon. Rafinesque G. 15 of N. G. American Animals, in Journal of Natural History Paris 1819.

This genus has been minutely described by Mr. Lesueur; yet it is strange that he should have hardly noticed the abdominal appendages, similar to those of the genera *Clupea, Salmo, Sparus,* &c. which are very large, acute flat scaly adipose, and on the external and lateral side of the base of each abdominal fin. This genus differs from *Clupea* and the foregoing genera by its mouth and teeth, abdomen and abdominal fins; it approximates also to *Erythrinus* and *Chirocentrus.* There are alreadyfive species known, all called Herrings on the Ohio: they appear early in the spring and disappear in the fall. They live on small fishes, insects, worms, and spawn: they often bite at the hook, and are taken in great quantities with the seines. I have adopted Mr. Lesueur's name, although it is not without objection, particularly by the similarity with *Diodon* in sound; but having divided the genus into three subgenera, one of the names given to them might, if needful, be adopted as the proper generic name.

1st Subgenus. AMPHIODON.

Body lanceolate, lower jaw longer, dorsal fin beginning opposite to the base of the anal fin. The name means toothed all over.

29th Species. TOOTHED FALSE HERRING. *Hyodon amphi-* *edon.* Do.

Amphiodon alosoides. Raf. 70 N. G. Animals. G. 15.

Diameter one fourth of total length, body silvery, back with bluish gilt shades, head gilt above: lateral line slightly curved downwards; tail acutely and equally forked, bluish brown, base reddish. Dorsal fin with 10 rays: anal fin with 34, ends acute; not falcated ~~————————~~

~~Length————————————————~~ conical acute ~~————————of the tongue.~~ Scales large ~~————————~~ ~~————the~~ mouth, round and black. Iris silvery gilt. Dorsal and anal fins with blue shades. It is very good to eat. I have observed it in the lower parts of the Ohio; where it is not so common as the two following species, and is often called Shad, owing to its larger size. Pectoral fins with 16 rays, and not reaching the abdominal fins. Tail with 24 rays.

~~30th Species.~~ ~~SWINEISH FALSE~~ HERRING, *Hyodon heteru-* ~~rus. Hyodon heterure.~~ ~~————————————~~

~~Diameter one fifth of total length, body~~ entirely silvery, oliva-ceous, brown above the head: lateral line straight raised up-wards at the base; tail acutely and ~~unequally forked,~~ the lower part longer. Dorsal fin with 12 rays, the anal ~~————————~~ cated, both ends obtuse.

Length from ten to twelve inches, body very narrow and com-pressed. Jaws with very small teeth; the lower jaw much lon-ger. Eyes over the corners of the mouth, ~~round and black, iris~~ gilt. Fins slightly olivaceous; the dorsal and anal have the two first rays simple ~~and the first very short, which~~ produce the obtuse appearance of those fins. Caudal with 24 rays, pectoral fins ~~with 16 rays and~~ reaching the abdominal fins. A common species in the Ohio and tributary streams; it appears later than the following; whence it is called Summer-herring. It forms a connecting link between this and the following subgenus, hav-ing the teeth as in the following species.

2d Subgenus. GLOSSODON.

Body lanceolate, jaws equal with small teeth; dorsal fin oppo-site to the vent, nearly medial, beginning behind the abdomi-nal fins. The name means toothed tongue.

31st Species. Spring False Herring. *Hyodon vernalis*, Hyodon printanier.

Diameter one fourth of total length, body entirely silvery, back with bluish shades: lateral line straight, tail equally forked, sinus obtuse. Dorsal fin with 13 rays, the anal with 28 rays, falcated and with acute ends.

Length from ten to twelve inches; head small and narrow, nostrils very large, eyes above the corner of the mouth, black and somewhat elliptical vertically, iris round, silvery with gilt shades. Fins slightly olivaceous, the dorsal with 3 simple rays, the first very short, anal fin somewhat adispose anteriorly. Pectoral fins with 12 rays, hardly reaching the abdominal fins. Tail with 30 rays, somewhat marginated with brown. Branchial membrane with 7 rays. This fish appears commonly over the Ohio and even at Pittsburgh in April; it is very common, but a poor food, owing to its great number of small bones. It is sometimes smoked and cured as the Atlantic Herrings; but is not quite so good.

3d Subgenus. Clodalus.

Body oblong irregular or somewhat rhomboidal. Jaws nearly equal, the lower one somewhat longer and with small teeth. Dorsal fin beginning before the base of the anal fin.

32d Species. May False Herring. *Hyodon clodalus*. Hyodon de May.

H. Clodalus. Lesueur Jour. Ac. N. Sc. 1. p. 377.

Diameter one fourth of total length, body silvery, back bluish, lateral line nearly straight, tail equally forked, sinus obtuse. Dorsal fin with 15 rays, the anal with 30, not falcated, ends acute.

Length eleven inches, fins yellow with metallic colours on the rays, pectoral with 13 rays not reaching the abdominal, caudal with 20 rays. It comes as far as Pittsburgh in May. Its flesh is pretty good. Eyes elliptical vertically, brown. Iris golden.

33d Species Lake False Herring. *Hyodon clodalus*, Hyodon lacustre.

H. tergisus. Lesueur Journ. Ac. N. Sc. 1. p. 336, tab. 14.

Diameter one fourth of total length, body silvery, back bluish, gill covers golden: lateral line somewhat flexuose or somewhat arched towards the back: tail equally forked, sinus obtuse. Dorsal fin with 15 rays, anal with 32, falcated, rounded anteriorly, acute behind,

This fish was observed by Mr. Lesueur in Lake Erie. Mr. Say thinks he has seen it at Pittsburgh; but I have never observed it in the Ohio, and I suspect that Mr. Say may have mistaken the *Hyodon* ~~verecundity for this species~~: in fact all the species ~~are not noticed by the fishermen and considered~~ as alike; ~~I have seen it~~ among the fishes of the Ohio ~~which~~ ~~it has~~ the same eyes and colours as the foregoing. Length thirteen inches. Good food. See Mr. Lesueur's minute description.

XIV Genus. Trout. Salmo. Truite.

Body somewhat cylindrical scaly, vent posterior. Gill cover double, scaleless, more than four rays at the branchial membrane. ~~Mouth large,~~ jaws with strong teeth. Two dorsal fins, ~~the first anterior or~~ opposed to the abdominal fins which have a scaly appendage, ~~the second adipose and~~ opposed to the anal fin.

This Linnean genus which includes the Trouts and Salmons is confined to the head waters and brooks ~~of~~ ~~~~ only know two species as yet; but there may be more in the small streams of Ohio, the Cumberland and Clinch mountains, &c. The white fish of Lake Erie, *Coregonus albus* of Lesueur, (or *Salmo clupeformis* of Dr. Mitchell,) a fish ~~which differs~~ from the Trouts by being toothless, ~~and is therefore a real Coregonus,~~ is said to be found in some streams of Indiana, at the head of the Wabash and Miami; but I have no certain proof of it. Other Trouts have been seen in the Osage river and other streams putting into the Missouri and Mississippi.

34th Species. Alleghany Trout. *Salmo Alleghanensis*. Truite alleganienne.

Back brownish, sides pale with crowded round fulvous spots, and some scattered scarlet dots above and beneath the lateral line, which is nearly straight: lower jaw hardly longer, tail reddish nearly lunulate, dorsal fin quadrangular with brown stripes, and ten rays: anal fin lanceolate whitish, with a longitudinal line black anteriorly and red posteriorly. .,

It is found in the brooks of the Alleghany mountains falling into the Alleghany and Monongahela. It has the manner of the small Brook-trouts, and is called Mountain-trout, Creek-trout, &c. It is easily caught with the hook, baited with earth-worms, and they may be enticed by rubbing the bait and hook with asafœtida like many other fishes. They afford a very good food. Length about eight inches. Head olivaceous with vio-let shades. Iris brown. Dorsal fin rufous with brown lines parallel with the back. Pectoral fins oval, not reaching the base of the dorsal nor abdominal fins, redish below, whitish a-bove, with a brown line. Abdominal fins with nine rays and similar to the pectoral fins in colour, scaly appendage very small. Tail with brown shades. Adipose fin acute. Diameter of the body one sixth of the total length. I have seen some individ-uals (they may be the female or a variety) who were of a paler colour, with fewer and smaller dots; they had the yellowish spots more crowded, the fins darker and the tail pale.

35th Species. BLACK TROUT. *Salmo nigrescens.* Truite noiratre.

Body blackish brown, with some small spots, head black; lat-eral line straight: lower jaw hardly longer; fins and tail black, tail slightly forked. Dorsal fin with 10 rays, anal fin with 15 rays.

A very rare species, seen only once, near the Laurel hills; it is said to be found also in the Yohogheny, a branch of the Mo-nongehela. Length six inches, diameter one fifth of total length. Iris black and gilt. Slightly pale under the body.

XV. Genus. MINNY. MINNILUS. Minny.

Body elongated, somewhat compressed, covered with small scales. Vent medial. Head flat above, and somewhat shielded. Gill cover double, scaleless, three branchial rays. Mouth diag-onal, small, toothless and beardless, without lips, lower jaw shorter and narrower. A small trapezoidal dorsal fin, nearer to the head than to the tail, opposite to the abdominal fins, and without spines. Abdominal fins with eight rays and without ap-pendages. (Tail forked in all the Ohio species.)

F

There are in the United States more than fifty species of small fresh water fishes, (and in the Ohio waters more than sixteen species) commonly called Minnies, Minnews, Bait-fish, Chubs, and Shiners, which should belong to the genus *Cyprinus* of Linneus, or rather to the part of it which has been called *Leuciscus* by Klein and Cuvier; which subgenus (or genus) is distinguished by a small dorsal fin, no spines nor beards; but as the genus *Cyprinus* forms now a large family, and even the genus *Leuciscus* must be divided, since it contains more than one hundred anomalous species, differing in the position of the dorsal fin and the vent, the number of rays to the abdominal fins, &c., I venture to propose this and the three following genera. Three other different genera might be established upon the European species, distinguished as follow:

Dobula. Dorsal fin nearer to the tail, abdominal fins with nine rays and an appendage: upper jaw longer.

Phoxinus differs by ten abdominal rays and no appendage.

Alburnus differs from Dobula by no appendage and the lower jaw longer.

Besides my genus *Hemiplus,* (Annals of nature,) which has the vent posterior, the lower jaw longer, only five rays and an appendage to the abdominal fins.

All these small fish are permanent; they feed on worms, insects, univalve shells, and spawn; they bite at a small hook, baited with worms or flies, and they form an excellent bait for all the larger fish which feed upon them. They are good to eat when fried.

56th Species. SLENDER MINNY. *Minulus dinemus.* Minny emeraude.

Diameter one eighth of total length, silvery, back olivaceous with a brown longitudinal stripe in the middle: two lateral lines, one straight, the lower curved downwards and shorter: head gilt and green above. Dorsal fin 9 rays. Anal fin 12 rays.
: A small and slender species, common in the Ohio, &c. and going in flocks; length two or three inches. Its head is beautiful when alive: it is above of a fine gold colour with green shades, becoming of an emerald green above the eyes. Iris silvery: sides opaque, upper lateral line gold-green. Nostrils

large. Pectoral fins with 12 rays, not reaching the abdominal. All the fins silvery. Tail with 24 rays. Scales very small.

37th Species. SPOTTED MINNY. *Minnilus notatus.* Minny tache.

Diameter one seventh of total length, silvery, back olivaceous with a large brown stripe in the middle; head brown above, lateral line straight, a black spot at the base of the tail. Dorsal with 8, and anal with 9 rays.

Same size with the preceding, but not so slender and less common. Iris golden, nostrils very large, mouth small, lateral line shining blue on the paque sides. Pectoral fins with 12 rays and not reaching the abdomen. Tail with 14 rays. It is often called Minny-chub.

38th Species. LITTLEMOUTHED MINNY. *Minnilus microstomus.* Minny microstome.

Diameter one seventh of total length, silvery, olivaceous on the back and head, sides with a few black dots: lateral line straight, pectoral fins reaching the abdominal fins. Dorsal and anal fins with eight rays.

A small species found in the Kentucky river. Mouth very small, nostrils large, iris silvery, fins fulvous, the pectoral with 12, and the caudal with 24 rays. Head elongated.

XVI Genus. SHINER. LUXILUS. Luxile.

Difference from *Minnulus.* Vent posterior or nearer to the tail. Mouth rather large, commonly with lips and equal jaws. Scales rather large. Preopercule with an angular suture.

1st Subgenus. CHROSOMUS.

Mouth large. without lips, lower jaw much shorter. Scales rough. Opercule flexuose.

39th Species. REDBELLY SHINER. *Luxilus erythrogaster.* Luxile erythrogastre.

Diameter one sixth of total length: back olivaceous brown, sides pale with two brown stripes, the upper reaching from the gills to the tail, and the lower from the nose to the anal fin; belly white with longitudinal red stripes from the pectoral fin to the tail: lateral line curved downwards and only anterior. Dorsal and anal fins elongated. Dorsal 8, and anal 7 rays.

A very distinct and insulated species, intermediate between

this and the foregoing genus. It might probably form a peculiar genus and be called *Chrosomus erythrogaster* or Kentucky Red belly. I saw it in the Kentucky river. Length from four to six inches. Tail forked as in all this family, and yellow as well as the dorsal fin, and with twenty rays. All the other fins are whitish. Head yellow above, silvery beneath, iris golden, the brown stripe goes across the eyes. Pectoral fins trapezoidal, with 12 rays, not reaching the abdominal fins. Lateral line reaching no further than the dorsal fin. Anal fin narrow. It is called Red belly Chub.

2d Subgenus. LUXILUS.

Mouth rather large, with small flat lips, jaws equal, scales large.

40th Species. GOLDHEAD SHINER. *Luxilus chrysocephalus.* Luxile chrysocephale.

Diameter one fifth of total length, silvery with golden shades on the sides, head gilt, back and nape dark olivaceous; lateral line curved downwards, pectoral fins reaching the abdominal. Dorsal and anal fins with nine rays.

Vulgar names, Gold Chub, Shiner, Goldhead, &c. Length six inches. It is found in the Kentucky, Ohio, Cumberland Green river, &c. Iris golden. Fins fulvous, the pectoral golden large with 14 rays: tail with 24. It resembles the common Shiner or Butterfish of Pennsylvania, *Cyprinus chrysoleucos* Mitchell; but that fish is a *Rutilus*, having nine abdominal rays, its body is besides shorter and the anal fin is falcated with 15 rays.

41st Species. KENTUCKIAN SHINER. *Luxilus Kentuckiensis.* Luxile du Kentuky.

Diameter one seventh of total length, silvery, back olivaceous, lateral line curved downwards, dorsal and caudal fins red, the pectoral yellow, not reaching the abdomen. Dorsal 8, and anal 7 rays.

Vulgar names, Indian Chub, Red tail, Shiner, &c. Length about four inches. It is reckoned an excellent bait for anglers, because it will swim a long while with the hook in its body. Eyes small, iris brown with a gold ring. Yellowish brown above the head. Abdominal and anal fins white. Pectoral and abdominal fins oboval, with 12 rays. Tail with 24 rays.

42d Species. YELLOW SHINER. *Luxilus interruptus.* Luxile jaunatre.

Diameter one sixth of total length: yellowish olivaceous above, silvery beneath, rufous brown above the head, a rufous line from the dorsal to the tail, two straight and separated half lateral lines, the anterior one above the posterior: pectoral fins reaching the abdominal. Dorsal with 10 and anal with 9 rays.

A small species, only three inches long, called Yellow Chub or Shiner. Seen in the Ohio. Sides opaque, with violet shades, Iris silvery, mouth large, lips very apparent. Fins yellowish, pectorals with 16 rays, caudals with 24.

XVII Genus. CHUBBY. SEMOTILUS. Semotile.

Difference from *Minnilus.* Vent posterior or nearer to the tail. Dorsal fin posterior, opposite to the vent and behind the abdominal fins. Mouth large and with lips. Scales rather large. Preopercule angular.

43d Species. BIGBACK CHUBBY. *Semotilus dorsalis.* Semotile dorsal.

Diameter one fifth of total length: silvery, back olivaceous with some black dots, and raised; head brown above, a crenulated keel above each eye: lateral line upwards at the base: tail brown, with a black spot at the base and another before it. Dorsal fin with 8 rays and a large brown spot at the anterior base. Anal fin with 9 rays.

It is found in the Kentucky, and several other rivers. Vulgar names, Big-back Minny or Chub, Skimback, &c. Length from three to six inches. Iris gilt brown. Fins olivaceous, pectoral fins with ie rays, trapezoidal not reaching the abdominal. Tail with 24 rays, and pale, base with a round black dot, and a smaller one before it on the body, when the lateral line terminates. Head separated from the back by a suture connected with the opercule, back large convex higher.

44th Species. BIGHEAD CHUBBY *Senotilus cephalus.* Senotile cephale.

Diameter one fifth of the total length: silvery, back brownish, lateral line raised upwards at the base; fins fulvous, the pectoral reddish, the caudal pale at the end and unspotted, the dorsal with nine rays and a large black spot at the anterior base, anal with nine rays.

Length from six to eight inches, not uncommon in the creeks of Kentucky, &c. Vulgar names Chub, Big-mouth, and Big-head. It has really the largest head and mouth of this tribe. Iris reddish iridescent. Pectoral fins with 15 rays trapezoidal and short, abdominal fins rounded, dorsal fin beginning over them, spot round. Tail with 20 rays.

45th Species. WARTY CHUBBY. *Semotilus diplemia.* Sem-otile verruqueux.

Diameter one sixth of total length: olivaceous brown with gold shades above, silvery beneath: lateral line double, the anterior and lower curved upwards at the base, reaching to the abdominal fins, the posterior and upper straight from the pectoral fins to the tail: fins red, a spot at the base of the dorsal and caudal, and many dots over them. Dorsal with nine rays; the anal with eight.

Length from three to four inches, often called Minny or Red-fin. Observed in the Kentucky river near Estill. The male fish has a larger mouth than the female and some black warts on the head. Fulvous brown on the head. Iris large, golden, and white. Some black dots on the dorsal and caudal fins: the caudal spot is on the tail, and the dorsal at the anterior base; they are small and round. The pectoral fins do not reach the abdominal fins; they have 18 rays: the tail has 24.

XVIII Genus. FALLFISH. RUTILUS. Rutile.

Difference from *Minnilus.* Vent posterior nearer to the tail. Abdominal fins with nine rays. Mouth large and with lips. Scales large.

I call this genus *Rutilus,* in the supposition that the *Cyprinus rutilus* may be the type of it; if it should be otherwise, it may be called *Plargyrus.*

46th Species. SILVERSIDE FALLFISH. *Rutilus plargyrus.* Rutile plargyre.

Diameter one fifth of total length: silvery, back with the dorsal, pectoral, and caudal fins olivaceous: lateral line curved downwards: snout truncate, mouth almost vertical. Dorsal and anal fins with nine rays.

Length from four to six inches: vulgar names, Silverside, Shiner, White Chub, &c. Common in the streams of Kentucky. Mouth large, upper jaw almost vertical, yet longer than the lower. Iris white. Pectoral fins with 14 rays, reaching almost the abdominals, which are oboval and white. Tail forked as usual with 24 rays. Scales large.

47th Species. BAITING FALLFISH. *Rutilus compressus.* Rutile appat.

Diameter one seventh of total length: silvery, back fulvous, sides compressed, lateral line straight, raised upwards at the base, snout rounded, mouth hardly diagonal, nearly horizontal. Dorsal and anal fins with nine rays.

A small fish from two to four inches long, called Fall-fish Bait-fish, Minny, &c. It is found in the Alleghany Mountains in the waters of the Monongahela, Kenhaway, and even in the Potomac. The name of Fall-fish arises from its being often found near falls and ripples. Body more compressed than in the other species, as much so as in the genus *Minnilus*. Scales large, lips a little fleshy. Iris silvery gilt. Fins transparent, the pectoral with 14 rays and not reaching the abdominal, tail with 32 rays.

48th Species. ROUNDNOSE FALLFISH. *Rutilus Amblops.* Rutile amblopse.

Diameter one sixth of total length: silvery, head fulvous above, snout round: sides with an opaque band, lateral line straight: pectoral fins with 12 rays and reaching the abdominal fins. Dorsal and anal fins with 10 rays.

Length one or two inches. Vulgar name White Chub or Fall-fish. It is found at the falls of the Ohio. Back slightly fulvescent, snout large and rounded, mouth hardly diagonal, eyes large, iris silvery, and scales large. Tail with 30 rays.

49th Species. BLACKTAIL FALLFISH. *Rutilus melanurus.* Rutile melanure.

Diameter one sixth of total length: silvery, back brownish: snout rounded, lateral line straight, tail blackish. Dorsal fin with 15 rays, anal with 12.

Length from four to six inches. Vulgar name Blacktail Chub. In the Ohio and Muskingum, &c. Head dark brown above, Mouth diagonal, iris silvery. Scales pretty large. Fins brownish, the lower ones pale, the pectoral short with 12 rays. Tail with 20 rays.

50th Species. ANOMAL FALLFISH. *Rutilus anomglus.* Rutile anomal.

Diameter one fifth of total length, fulvous above, sides dusky, white beneath: snout rounded, a vertical brown line behind the gills; lateral line straight raised upwards at the base: pectoral fins yellow oboval short with 15 rays: tail unequally bilobed, the upper lobe larger. Dorsal and anal fins red, dorsal 8 and anal 7 rays.

An anamalous fish, differing from all those of the Cyprinian tribe in the Ohio, by its unequal bilobed tail, which is brownish and has 22 rays. Mouth diagonal. Eyes small, iris olivaceous gilt. Nape of the neck red, scales rather small. Length three inches. Found in Licking river &c. Vulgar names Chub, Redfish, Fallfish, &c.

51st Species. RED MINNY. *Rutilus? ruber.* Rutile rouge. Entirely red, tail forked.

I add here a fine small fish, which I have never as yet, but is said to live in the small streams which fall into the Elkhorn and Kentucky. It is a slender fish, only two inches long, compressed and of a fine purple red. It may belong to this genus, or to any other of this tribe. It is commonly called Red-minny.

XIX Genus. FLAT-HEAD. PIMEPHALES. Pimephale.

Body oblong, thick, and scaly. Vent posterior nearer to the tail. Head scaleless, fleshy all over, even over the gill covers' rounded, convex above and short. Mouth terminal small, toothless, with hard cartilaginous lips. Opercule double, three branchial rays. Nostrils simple. Dorsal fin opposite the abdominals, with the first ray simple and cartilaginous. Abdominal fins with eight rays.

A singular new genus, which differs from *Catostomus* by the terminal mouth, hard lips, soft head, simple dorsal ray, &c. The name is abbreviated from Pimelecephales which means Flat-head.

52d Species. BLACKHEADED FAT-HEAD. *Pimephales pro-melas.* Pimephale tete-noire.

Diameter one fourth of the length, body olivaceous silvery, head blackish, snout truncated, and with soft warts: fins whitish, dorsal with a large irregular black spot at the anterior base, with eight forked rays, and one simple shorter obtuse hard: anal with eight rays; lateral line flexuose and raised at the base, tail lunulate.

A small fish three inches long. It is rare and hardly known by the anglers. I describe it from a specimen taken with a hook baited with earth-worm, by Mr. William M. Clifford, in a pond near Lexington, in the month of April 1820, and now preserved in the Museum in Lexington. Its head is very remarkable, soft and fat all over, the snout sloping, broad, truncate with soft warts in front, mouth at its inferior extremity very small, elliptical transversal, with equal circular hard lips. The whole head and even the eyes are of dusky and bluish black colour. Pectoral fins trapezoidal with 15 rays, the upper rays of the colour of the head. Tail olivaceous lunulated, with 20 forked rays and 5 short simple rays on each side of the base. Abdominal fins quadrangular. The first ray of the dorsal is singular, thick, short, hard, and yet blunt, almost cartilaginous, or not properly spinous, and not at all serrate as in the Carps. Scales pretty large.

XX GENUS. SUCKER. CATOSTOMUS. Catostome.

Body oblong cylindrical scaly. Vent posterior or nearer to the tail. Head and opercules scaleless and smooth. Mouth beneath the snout, with fleshy, thick, or lobed sucking lips: Jaws toothless and retractible. Throat with pectinated teeth. Nostrils double. Gill-cover double or triple. Three branchial rays to the gill membrane. A single dorsal fin commonly opposite to the abdominal fins, which has from eight to ten rays.

Lesueur has established this genus, in the first volume of the Journal of the Academy of Natural Sciences of Philadelphia, with all the American species of the genus *Cyprinus* which have the above characters, and he has described eighteen species belonging to it. I have discovered twelve additional new species in the waters of the Ohio, where about sixteen new spe-

cies have already been detected. This genus having become
so extensive at an early period, and many other species existing
probably in North America and Siberia, I have therefore divided it into five subgenera, employing principally the number of
abdominal rays. All these fishes are permanent in the Ohio
its branches and the ponds. Some however disappear in winter,retreating into deep water or into the mud. Many bite at the
hook. They feed on univalve shells, small fishes and spawn.
They offer a tolerable food.

             1st. Subgenus. Moxostoma.

" Body oblong, compressed; head compressed, eight abdominal
rays, dorsal fin commonly longitudinal, tail commonly unequally forked.

53d Species. Ohio Carp Sucker. *Catostomus anisurus.*
Catostome anisure.

Diameter one fifth of the length: silvery, slightly fulvescent
above, fins red, the dorsal olivaceous falcated with 17 rays,
nearer to the head and reaching the vent: lateral line curved upwards and flexuose at the base: snout gibbose: tail forked,
upper part longer. Anal fin falcate with eight rays.

A large species common all over the Ohio and the large
streams, as far as Pittsburgh. Permanent and sometimes taken
in winter. It is called Carp every where. Length from one to
three feet. It is taken with the hook, seine, and dart. Its
flesh is pretty good, but soft. The male fish has a red tail;
while it is olivaceous in the female. Snout divided from the
head by a transverse hollow which makes it gibbose. Eyes
black, iris silvery and golden above. Sides often with copper
shades. Scales large with concentric stria. Pectoral fins large
oval acute with 15 rays and reaching the abdominal fins. Caudal with 24 rays.

54th Species Buffalo Carp Sucker. *Cotostomus anisop-*
turus. Catostome anisopture.

Diameter one fourth of the total length: silvery: head sloping, lateral line curved as the back: tail unequally bifid, upper
part much longer: dorsal fin longitudinal, beginning above the
pectorals and reaching the end of the anal, sinuated by a double falcation, first ray very long.

A singular species which I have never seen. I describe it from a drawing of Mr. Audubon. It is found in the lower part of the Ohio, and is called Buffalo Carp, Buffalo perch, Buffalo Sucker, White Buffalo-fish, &c. Length about one foot. Very good to eat. Taken with the seine in the spring only. Body broad, dorsal fin broad and large, remarkable by its shape like a double sickle, and first ray which reaches the tail. Anal fin small and falcate. Pectoral fins reaching the abdominal fins. The number of abdominal rays was not observed, if it should have nine it would be nearer to *C. Velifer* and *C. seto-sns*, or it may form a peculiar subgenus.

The *C. tuberculatus* of Lesueur belongs also to this subgenus, having eight abdominal rays; but its tail is regularly bifid.

2d Subgenus. ICTIOBUS.

Body nearly cylindrical. Dorsal fin elongated, abdominal fins with nine rays, tail bilobed, commonly equal.

The *C. gibbosus* and *C. Communis,* of Lesueur, appear to be intermediate between this subgenus and the foregoing, having nine abdominal rays, but an unequal bilobed tail.

55th Species. BROWN BUFFALO-FISH *Catostomus bubalus.* Catostome bubale.

Diameter one fifth of the total length; olivaceous brown, pale beneath; fins blackish, pectoral fins brown and short: head sloping, snout rounded, cheeks whitish: lateral line straight, dorsal fin narrow with 28 equal rays, anal trapezoidal with 12 rays.

One of the finest fishes of the Ohio, common also in the Mississippi, Missouri, and their tributary streams. It is called every where Buffalo-fish, and Pizoneau, by the French settlers of Louisiana. I had called it *Amblodon bubalus* in my 70 N. G. of American Animals, having been misled by the common mistake which ascribed to it the teeth of the *Amblodon grunniens*; but it is a real *Catostomus,* without any such teeth. Length from two to three feet; some have been taken weighing thirty pounds and upwards. It is commonly taken with the dart at night when asleep, or in the seine; it does not readily bite at the hook. It feeds on smaller fishes and shells, and often goes in shoals. It retires into deep water in the winter, yet is sometimes taken even then. It comes as far as Pittsburgh. Its flesh

is pretty good but soft. Scales rather large. Tail with 24 rays
and two equal rounded lobes. Iris gilt brown, eyes small. Pectoral fins with 16 rays. Dorsal fins shallow and even beginning
just before the abdominal fins, and ending at the base of the a-
nal fin.

56th Species. BLACK BUFFALO-FISH. *Catostomus niger.*
Catostome noir.

Entirely black, lateral line straight.
I have not seen this fish. Mr. Audubon describes it as a peculiar species, found in the Mississippi and the lower part of the
Ohio, being entirely similar to the common Buffalo-fish, but
larger, weighing sometimes upwards of fifty pounds, and living
in separate shoals.

3d Subgenus. CARPIODES.
Body oblong, somewhat compressed; head compressed, nine
abdominal rays, dorsal fin commonly elongate,tail equally forked.
The *C. cyprinus* and *C. setosus*, of Lesueur, belong to this
Subgenus.

57th Species. OLIVE CARP SUCKER. *Catostomus carpio.*
Catostome carpe.

Diameter one fourth of the length: olivaceous above, pale beneath, chin white, abdomen bluish: lateral line straight, dorsal
fin somewhat falcated with 36 rays, anal trapezoidal with 10
rays; head sloping, snout rounded.

Seen at the falls of the Ohio, commonly called Carp. Length
from one to two feet. Eyes very small and black, fins olivaceous brown, the pectorals olivaceous, trapezoidal short and
with 16 rays. Tail with 24. Dorsal fin beginning before the
abdominal and reaching the end of the anal fin. Not so good to
eat as the Buffalo-fish.

58th Species. SAILING SUCKER. *Catostomus velifer.* Catostome volant.

Diameter less than one fourth of the length: body elliptical,
silvery with golden shades, lateral line flexuose, dorsal fin very
broad falcated with 25 rays, the first ones very long, anal fin trapezoidal lunulate with 10 rays: head sloping, snout rounded.

Catostomus anonymous Lesueur in Journ. Ac. Nat. Sc. of
Philadelphia, Vol. 1, page 110.

A singular fish, not very common, yet found as far as Pittsburgh. It has received the vulgar names of Sailor fish, **Flying**
fish, and Skimback, because, when it swims, its large dorsal fin
appears like a sail, and it often jumps or flies over the water
for a short distance. Length commonly from twelve to sixteen
inches, of which the tail, which is very large, includes one
fourth, and has 24 rays. Back slightly olivaceous, scales very
large. Fins olivaceous brown, except the abdominal and pectoral, which are white. The dorsal beginning before the abdominal and reaching the end of the anal, the second and third
rays are one third of the whole body, the first is short and cleav-
ing to the second; mouth small, quite terminal at the lower end
of the rounded snout; head small, convex above. Pectoral fins
with 16 rays. Not very good to cat. Seen only in summer.

59th Species. Mud Sucker. *Catostomus xanthopus.* Catostome xanthope.

Diameter one fourth of the length: lateral line straight: silvery, back olivaceous, head brown above, snout gibbose rounded: dorsal fin hardly falcate with 14 rays, anal lanceolate with
8 rays: lower fins yellowish.

Found below the falls. Length from six to ten inches. It
lives in muddy banks, and conceals itself in the mud. Flesh very soft. Head large, flattened above, mouth large, eyes large.
Iris silvery. Lateral line hardly raised at the base. Dorsal fin
above the abdominal, fins olivaceous as well as the tail, which
has 20 rays. Pectorals with 18 rays. Scales large.

4th Subgenus. Teretules.

Body elongate cylindrical or somewhat quadrangular, nine
abdominal rays, dorsal fins commonly small, tail equally forked.

An extensive Subgenus to which belong all the following
species of Lesueur: *C. aureolus, C. macrolepidotus, C. longi-
rostrum, C. nigricans, C. vittatus, C. maculosus, C. Sucetta,*
besides the *C. teres* and *C. oblongus* of Mitchell.

60th Species. Black-face Sucker. *Cotostomus melanops.*
Catostome melanopse.

Diameter one seventh of the length: head squared, blackish
above, snout convex obtuse; back olivaceous, sides whitish
with scattered black dots, a black spot on the gill cover, and a

large one between the dorsal and caudal fins: lateral line straight, dorsal fin with 14 rays, anal with 9 rays.

A singular species seen at the falls. It is rare and called Spotted Sucker or Black Sucker. Length from four to six inches; body cylindrical, flattened beneath as far as the vent. Head flat above, blackish there and in the fore part. Mouth almost terminal with thick whitish lips, the lower one shorter and thicker, a few small black spots on the sides of the head and a large one on the preopercule. Gill cover silvery. Eyes black, iris brown with a gold ring. Back of a rufescent colour with gold shades. A very large black patch above the anal fin before the tail. Sides pale with small unequal black dots, belly whitish: Fins coppery, the pectoral elliptical elongated with 18 rays, the anal elongated reaching the tail, the dorsal broad and opposed to the abdominal. Tail with 20 rays. Scales rather large nervose radiated.

61st Species. BLACK-BACK SUCKER. *Catostomus melanotus.* Catostome melanote.

Diameter one sixth of the length: bluish black above, whitish beneath; head convex, snout obtuse: lateral line straight: dorsal and anal fins with nine rays.

Seen only once at the falls. Length six inches, body nearly cylindrical. Mouth rather inferior, lips thick and somewhat gristly. Iris silvery. Scales pretty large. Fins whitish, the dorsal and caudal a little redish. Pectoral fins elliptical with 16 rays. Tail 20. Dorsal fin trapezoidal, opposed to the abdominal, the first ray shorter. Anal elliptical obtuse. Vulgar names Black Sucker and Blue Sucker.

62d Species. ROUGH-HEAD SUCKER. *Catostomus fasciolaris.* Catostomus fascie.

Diameter one sixth of the length: brown above, white beneath, sides with small transversal black lines: head sloping, tuberculated above, snout obtuse: dorsal fin longitudinal reaching the end of the anal fin, lateral line straight.

I have not seen this species, but describe it from a drawing of Mr. Audubon. It is found in the lower part of the Ohio. Vulgar names Rough-head Sucker, Pike Sucker, Striped Sucker. Length about eight inches, body cylindrical tapering behind,

Eyes small, mouth beneath. Lower fins trapezoidal, about twenty transversal lines. A doubtful species, perhaps an Hydrargyrus, but the mouth is like that of the Sucker.

63d Species, RED-TAIL SUCKER. *Catostomus erythrurus.* Catostome rougequeue.

Diameter one fifth of the length: rufous brown above, white beneath; tail olivaceous: head convex, snout rounded: lateral line straight: dorsal fin trapezoidal redish with 12 rays. anal fin elongated yellow, anal falcated, with 7 rays.

A fine species, not uncommon in the Ohio, Kentucky, Cumberland, Tennessee, &c. Vulgar names Red-horse, Red-tail, Horse-fish, Horse Sucker, &c. Length about one foot. Scales very large. Mouth beneath. Iris whitish, eyes black. Pectoral fins yellow elliptical reaching the abdominals and with 16 rays. Tail large with 20 rays. Its flesh is dry and not very good to eat.

64th Species. KENTUCKY SUCKER. *Catostomus flexuosus.* Catostome flexueux.

Diameter one fifth of the length: silvery, back brownish, scales rather rough, opercule flexuose: head squared, snout gibbose truncate; lips very thick, the inferior bilobed: lateral line flexuose; tail brown: dorsal fin blackish with 12 rays, anal fin whitish with 7 rays and reaching the tail.

The most common species in Kentucky, in all the streams and ponds, called merely Sucker. Very good to eat. It conceals itself in the mud in winter. It bites at the hook, living on minnies and little lobsters. Body thick cylindrical. From ten to twelve inches long. Head large, a deep depresion between the snout and the head, mouth large with fleshy lips. Eyes large black, iris yellow. Opercule hard bony. Lower fins whitish, pectorals elongated elliptical with 20 rays. Tail 20 rays. Dorsal trapezoidal sloping behind. This fish is the most useful to keep in ponds.

65th Species. BIG-MOUTH SUCKER. *Catostomus? megastomus.* Catostome megastome.

Diameter one fifth of the length: blackish above, yellowish beneath, very broad: a spine at the base of the pectoral fins: lateral line straight.

A very doubtful species seen by Mr. Audubon. It comes sometimes in shoals in March, and soon disappears. Only taken with the seine, not biting at the hook; vulgar name Brown Sucker. The mouth is very remarkable, being broader than the head, somewhat projecting on the sides. Length one foot. The head resembles that of Cat-fish, but has no barbs. Is it a peculiar genus owing to the mouth and pectoral spine? It might be called *Euryostomus*. The yellow colour covers the forehead and reaches to the anal fin. Dorsal opposed to the abdominal and trapezoidal, pectorals elliptical yellow.

5th Subgenus. DECACTYLUS.

Body nearly cylindrical, abdominal fins with 10 rays: tail equally forked.

Besides the two following species, the *C. bostoniensis* and *C. hudsonius*, must be enumerated here.

66th. Species. PITTSBURGH SUCKER. *Catostomus duquesni.* Catostome duquesne.

Diameter one fifth of the length, whitish; lateral line curved towards the back: anal fin with nine rays extending to the tail: dorsal with 14 rays and trapezoidal.

C. duquesni Lesueur J. Ac. Nat. Sc. v. 1, p. 105.

This species has been pretty well described by Lesueur his description. Length from 15 to 20 inches: good to eat, found in the Ohio as far as Pittsburgh: vulgar name White Sucker.

67th Species. LONG SUCKER. *Catostomus elongatus.* Catostomus allonge.

Diameter one seventh of the length; brownish; lateral line nearly straight: snout and opercules tuberculated: dorsal fin with 33 rays, long, falciform and raised anteriorly. Anal fin small with 8 rays.

C. elongatus Lesueur J. Ac. Nat. Sc. v. 1, page 103.

It is found in the Ohio as far as Pittsburgh, and called Brown Sucker, length from 20 to 25 inches. Head small cuneiform above: Scales large. Good to eat. See Mr. Lesueur's description.

XXI Genus. Suckrel. Cycleptus. Cyclepte.

Difference from the foregoing genus—Two dorsal fins, mouth round and terminal.

The name means *small round* mouth.

63th Species. Black Suckrel. *Cycleptus nigrescens.* Cyclepte noiratre.

Blackish, belly whitish, mouth recurved, tail forked. *Cycleptus.* 17th G. of Prod. 70 N. G. American Animals. A singular and rare fish, which I have never seen, but mention upon the authority of Mr. Bollman of Pittsburgh; where it sometimes appears in the spring; but it is a rare fish, whose flesh is very much esteemed. It is also found in the Missouri, whence it is sometimes called the **Missouri Sucker.** Length two feet.

XXII Genus. Catfish. Pimelodus. Pimelode.

Body scaleless, elongated. Head large with barbs. Two dorsal fins, the second adipose and separated from the tail, the first short and commonly armed. Pectoral fins commonly armed. Teeth like a file. Vent commonly posterior.

The extensive genus *Silurus* of Linneus, which is scattered throughout the rivers of both continents, has not yet been completely illustrated, notwithstanding the labours of the modern icthyologists: I have found in the Ohio about twelve species belonging to it: most of which offer consimilar characters and appear to belong to the genus *Pimelodus* of Lacepede and Cuvier: which have left the name of *Silurus* to the species having one dorsal fin. I have already published a monography of them in the Journal of the Royal Institution of London, under the generic name of *Silurus.* I now propose to form with them a peculiar *subgenus*, divided in many sections, and different from the subgenera *Bagrus, Synodontus, Silusox,* &c.

Subgenus. Ictalurus.

Head depressed with eight barbs, one at each corner of the mouth, longer than the others, four under the chin, and two on the snout behind the nostrils. Teeth in two patches, acute and file-shaped. Pectoral fins and first dorsal fin armed with an anterior spine. First dorsal trapezoidal and before the abdomi-

H

nals, second opposite the anal. Body compressed behind, vent posterior or sub medial. Operculum simple.

The fishes belonging to this group are common throughout the United States, the *Silurus catus* of Linneus, which is not found in the Ohio, belongs also to it. They are sedentary in the Ohio and branches, and very voracious, feeding on all smaller fishes: they are easily taken with the hook; their flesh is esteemed, and, although it is somewhat tough in the largest species, it makes notwithstanding excellent soup. These fishes often come to a great size, and live to a great age. The name of *Ictalurus*, means Cat-fish in Greek.

1st Section. ELLIOPS. Tail forked. Eyes elliptical. Abdominal fins with less than nine rays.

69th Species. SPOTTED CATFISH. *Pimelodus maculatus,* Pimelode tachete.

Upper jaw longer, lateral barbs black, reaching the dorsal fin. Eyes elliptical. Body whitish with small unequal brown spots on the sides; vent submedial: tail unequally forked, upper lobe longer. Pectoral fins fenestrated. Anal fin longitudinal with 27 rays. Lateral line straight.

Silurus maculatus. Monogr. sp. 1.

One of the small species, commonly about one foot long and slender, never reaching a large size. Vulgar names Spotted, White, and Channel Catfish. It is found as far as Pittsburgh, but is not very common. Flesh very good. Head long and flat, olivaceous rufous above, jaws rounded, lips thick. Upper barbs the shortest and white; the exterior inferior ones long and black at the end. Iris elliptical white. Body somewhat attenuated behind, entirely silvery white. Belly white, flattened, without spots or shades. Sides with gilt and blue shades, besides the brown spots. Back unspotted, pale, rufescent. Lateral line not reaching the gills and slightly raised upwards at the base. First dorsal fin with six soft rays. Pectoral fins with five, spiny ray longer, very thick, and united to the fin by a fenestrate web on the inner serrate side. Abdominal oboval and with 8 rays. Caudal with 20. Lobes acute. All the fins redish, marginated, or tipped with brown. Tail marginated, Adipose fins brown.

70th Species. BLUE CATFISH. *Pivelodus cerulescens*. Pimelode bleuatre.

Upper jaw longer, lateral barbs black, shorter than the gills. Eyes elliptical. Operculum and lateral line flexuose. Body of a bluish lead colour, whitish beneath, unspotted. Tail equally forked, base redish. Anal fin arched with 25 rays, *Silurus cerulescens*. Monogr. sp. 3.

...A fine species, reaching sometimes to a very large size, I have been told that one was taken weighing 185 pounds and another 250 pounds. Vulgar names Blue Cat and Brown Cat, or Catfish. It is not uncommon in the lowest parts of the river. Whole shape somewhat fusiform as in all the species with a forked tail, yet depressed forwards and compressed behind. Of an uniform lead colour, nearly blue in the young individuals and nearly brown in the old ones. Barbs rather short and white, the upper ones very short and brown. Iris elongate and whitish. Fins bluish; but the pectoral and abdominal whitish. Spine of the pectoral fins equal in length, not fenestrate, and hardly serrate inside. Number of rays, dorsal 1 and 6, pectoral 1 and 7, abdominal 6, caudal 22. A variety has a blackish tail. Vent posterior.

71st Species. WHITE CATFISH. *Pimelodus pallidus*. Pimelode pale.

Upper jaw longer, lateral barbs reaching the pectoral fins. Eyes elliptical. Lateral line straight. Body whitish, back slightly olivaceous. Tail nearly equally forked. Anal fin elongate with 25 rays.

Silurus pallidus. Monogr. sp. 2.

Vulgar names white and channel Catfish: this last name is given to it because it dwells principally in the channels or deeper parts of the river. Length from one two to feet. Shape as in the foregoing. Head smaller, olivaceous above. Barbs white. Iris white. First dorsal fin nearer to the abdominal fins, yellowish, rays 1 and 6. Pectorals yellowish, rays 1 and 7. Abdominals white with six rays. Adipose fin olive with a brown tip. Anal and caudal pale brown, 24 rays in the tail, which has the upper acute lobe slightly longer. It offers some varieties. 1st. *Marginata*. Tail fulvous, marginated with black

2d. *Lateralis.* With three black spots on each side. 3d. *Leu-cofitera.* All the fins pale and whitish.

72d Species. SILVERY CATFISH. *Pimelodus argyrus.* Pimelode argyre.

' Jaws nearly equal, lateral barbs brown and reaching the pectoral fins. Eyes elliptical. Body silvery, lateral line straight. Fins brownish, anal with 25 rays. Tail equally forked.

Silurus argenteus. Monography, sp. 4. There is another species of that name already.

A small and rare species, very similar to the foregoing, of which it is perhaps a variety. Number of rays similar.

2d Section. LEPTOPS. Tail bilobed. Eyes round and very small. Nine abdominal rays. Vent posterior. Adipose fins large.

73d Species. CLAMMY CATFISH. *Pimelodus viscosus.* Pimelode visqueux.

Jaws nearly equal, barbs very short, eyes round over the head. Body entirely brown, lateral line raised upwards before. Pectoral fins with 1 and 7 rays, anal fin rounded with 15 rays. Tail unequally bilobed and black; upper lobe smaller and white.

Silurus viscosus. Monogr. sp. 6

A very singular and rare species, found at the full length only 4 inches. brown with bluish and greyish shades, covered with a clammy viscosity; throat whitish. Head very flat, with a longitudinal furrow above, elongated; upper jaw hardly longer. Eyes over the head very small and bluish. Spines of the anterior fins short, thick, and simple. Dorsal with 1 and 7 rays. Abdominal small with 9. Anal blackish.

75th Species. CLOUDED CATFISH. *Pimelodus nebulosus.* Pimelode nebuleux.

Jaws equal, barbs shorter than the head. Eyes round, exceedingly small. Body olivaceous, clouded with pale brown, white beneath, lateral line nearly straight. Pectoral fins with 1 and 9 rays, anal fin rounded with 12 rays. Tail merely notched, hardly but equally bilobed.

Silurus nebulosus. Monogr. sp. 5.

This species is totally different from the foregoing, and might perhaps form a peculiar section or even subgenus, (O-

pladelus,) by the conical head, membranaceous operculum; but particularly because the first ray of all the fins, except the caudal and adipose, is a kind of soft obtuse spine concealed under the fleshy cover of the fins. It is a large fish, from two to four feet long, and commonly called Yellow Cat, Mud Cat, and Brown Cat; but these names are common to other species. It is very good to eat, either boiled or fried. Head conical depressed, iris redish brown, eyes black, lateral barbs white, the lateral ones brownish. Operculum with a large membranaceous appendage or flap. Body conical tapering behind. Dorsal fins with 1 and 6 rays. All the fins very fat, thick, and somewhat redish, abdominal fins brownish. Tail with 20 rays.

2d Section. AMEIURUS. Tail entire. Eyes round. Eight abdominal rays. Vent posterior. Dorsal fin anterior with a spine. Lower jaw not longer. Pectoral fins, with one simple spine and seven rays.

75th Species. YELLOW CATFISH, *Pimelodus cupreus.* Pimelode cuivre.

Upper jaw longer, barbs half the length of the head. Eyes round. Body entirely of a coppery yellow colour. Lateral line straight. Tail truncate entire. Anal with 15 rays.

Silurus cupreus. Monogr. sp. 9.

Vulgar name, Yellow Catfish. - Very different from the foregoing. Similar however in size and form. Colour uniform, extending on the head and fins. Spines short. It is found as far as Pittsburgh. Very good to eat. Some have been taken weighing over 200 pounds. Dorsal fin with 1 and 7 rays.

76th Species. BROWN CATFISH, *Pimelodus lividus.* Pimelode livide.

Jaws equal, barbs nearly equal together and as long as the head. Eyes round. Body entirely of a livid brown colour. Tail rounded entire. Lateral line raised upwards at the base. Anal fin elongate with 25 rays.

Silurus lividus. Monogr. sp. 7.

A small species, entirely of a leaden brown. Head short, slightly olivaceous, throat pale. Barbs equal, the upper ones livid, the lower ones rufous. A furrow on the head which is convex above. Operculum flexuose. Tail with 24 rays. Dorsal with one and 7. Spines short.

77th Species. BLACK CATFISH. *Pimelodus melas.* Pimelode noir.

Jaws nearly equal. Eyes round. Barbs unequal, shorter than the head. Body entirely black, lateral line straight. Anal fin with 20 rays. Tail nearly truncate, entire.

Silurus melas. Monogr. sp. 8.

A rare species less than a foot long. Hardly pale beneath. Dorsal fin 1 and 7. Found below the falls.

Species. YELLOW HEAD CATFISH. *Pimelodus xanthocephalus.* Pimelode xanthocephale.

Upper jaw longer. Barbs unequal shorter than the head. Eyes round. Body iron grey, with the whole or part of the head yellow. Belly white. Lateral line straight. Anal fin with 22 rays. Tail entirely truncate.

Silurus xanthrocephalus. Monogr. sp. 10.

About a foot long. In the Ohio, Kentucky, &c. Head very large, often entirely yellow, or only forward, or covered with yellow patches. Iris white. Fins fleshy redish. The dorsal with 1 and 6 rays, caudal 24. Good food.

4th Section. ILICTIS. Tail entire, eyes elliptical abdominal rays. Dorsal fins submedial. Pectoral fins with one flat spine serrated outwards, and nine rays. Lower jaw longer.

79th Species. MUD CATFISH. *Pimelodus limosus.* Pimelode bourbeux.

Lower jaw longer. Barbs black, the lateral ones reaching the pectoral fins. Body fulvous, variegated or clouded with black, belly grey. No lateral line. Anal fin with 15 rays. Tail entire oval obtuse.

Silurus limosus. Monog. sp. 11.

A very singular species, differing from all others by the long lower jaw, &c. Length about one foot. It has a slender body of a rufous brown mixed with black. It is found in the muddy streams, and near the muddy banks of large rivers. Dorsal fin opposite the abdominal, with one spine concealed under the skin and six rays. Branchial membrane apparent outside. Pectoral fins with 10 rays, the first whereof is a long and broad flat spine, barbed outwards. Tail with 20 rays. This fish can live very long out of water, and is sometimes alive 24 hours after having been taken.

XXIII Genus. Mudcat. Pilodictis. Pylodicte.

Body scaleless conical flattened forwards and compressed behind. Head very broad and flat, with barbs, eyes above the head. Two dorsal fins, both with soft rays. Vent posterior. This genus was the 10th of my Prod. of 70 N. G. of Animals. The name means Mudfish. It differs principally from the foregoing by the second dorsal having rays.

80th Species. Toad Mudcat. *Pylodictis limosus.* Pylodicte bourbeux.

Lower jaw longer, eyes round, eight barbs, four above and four below. Head verrucose above. Body brown, clouded and dotted with yellowish, redish, and bluish, one row of transversal black lines on each side of the back. No lateral line. Tail entire and truncate.

I have not seen this fish, but describe it from a drawing of Mr. Audubon. In is found in the lower parts of the Ohio and in the Mississippi, where it lives on muddy bottoms, and buries itself in the mud in the winter. It reaches sometimes the weight of 80 pounds. It bears the name of Mudcat, Mudfish, Mudsucker, and Toadfish. It is good to eat and bites at the hook. The head is broader than the body and with a very large mouth; the barbs appear to lay in four pairs, two above, longer and near the nostrils, and two smaller under the lower jaw. The first dorsal fins triangular and above the abdominals, which are nearer the pectorals than to the anal. Second elongate with many rays. Number of rays unnoticed.

XXIV Genus. Backtail. Noturus Noture.

Difference from G. *Pimelodus*, S. G. *Ictalurus*, and Sect. *Amejurus*: Adipose dorsal fin very long, decurrent and united with the tail, which is decurrent on each side, but unconnected with the anal fin.

Genus 18th of the Prodr. N. G. It differs from the genus *Plotosus* of Lacepede by having the anal fin free, and from *Pimelodus* by the connection of the tail with the second dorsal fin. The name means Tail over the back. The *Silurus gyrinus* of Mitchell must belong to this genus.

81st Species. YELLOW BACKTAIL. *Noturus flavus.* Noture jaune.

Entirely yellowish. Upper jaw longer, barbs half the length of the head. Eyes round. Lateral line nearly straight. Anal fin with 14 rays. Tail entire truncate.

A small species very common near the falls. Length 4 to 12 inches. It agrees in almost every thing with the Section *Ainalura* among the Catfihes. Vulgar name Yellow Catfish, like the *Pimelodus cupreus.* Dorsal fin with 1 and 7 rays, rounded spine very short and obtuse. Second dorsal beginning before the anal and extending to the tail in a curve. All the lower fins rounded. Pectorals with 1 and 7 rays, spine equal and acute. Abdominal fins with 8 rays. All the fins fleshy and fat. Head flat above, barbs unequal. Belly convex. Hind part of the body compressed.

XXV Genus. TOTER. HYPENTELIUM. Hypentele. Body pyramidal slightly compressed, with very minute scales. Vent posterior. Head scaleless nearly square, mouth terminal protruded beneath toothless, lower jaw shorter with five lobes, the middle one larger, lips very small. Abdominal fins anterior, removed from the vent, with nine rays, dorsal fin anterior opposed to them.

This genus belongs to the family of Cyprinidia, and is next to my genus *Exoglossum*, with which I had united it; but this last differs from it by an oblong body, flat head, lower lip trilobe not protruded, abdominal fins and dorsal fin medial, &c. The name expresses the character of the lower lip.

82d Species. OHIO TOTER. *Hypentelium macropterum.* Hypentele macroptere.

Forehead sloping truncate tuberculated. Body silvered, variegated, and reticulated with blackish, lateral line straight and faint. All the lower fins elongated, the pectorals reaching the abdominals, the anal with 10 rays and reaching the tail. dorsal fin with 12 rays, tail forked.

Exoglossum macropterum. Raf. in Journal Acad. Nat. Sc. of Philad. Vol. 1, page 320. tab. 17. fig. 4.

It is found near the falls and is only a small fish 2 or 3 inches long. Its vulgar name is Toter or Stone Toter. (Toter is a Virginia name for carrier.) There is a kind of Chub in Virginia which bears the same name and has the habit of pushing pebbles with its head in order to form an inclosure where the female lays its eggs; the name of Toter was given to the Ohio fish owing to the same peculiarity. It is a rare fish and used as bait. The mouth projects in a short and obtuse snout. Iris large and gilt. Opercule simple. Pectoral fins lanceolate acute, as long as the head and with 12 rays. Abdominal fins lanceolate acute, situated nearly half way between the head and the vent, but not reaching it. Dorsal fin trapezoidal. Anal fin elongate. Caudal with 20 rays.

XXVI Genus. RIBBONFISH. SARCHIRUS. Sarchire.
Body scaleless slender cylindrical, slightly compressed. Vent posterior. Head nearly square. Jaws elongated narrow flat, with four rows of small unequal teeth, the lower one shorter and moveable, the upper one longer immobile, with an obtuse knob atthe end. Pectoral fins round without rays, but with a thin circular membrane surrounding an adipose base. Abdominal fins anterior with six rays. Dorsal fin posterior nearer to the tail than the anal. Caudal fin lanceolate, decurrent beneath.

A very distinct genus of the family Esoxida, differing from all the genera of it by its fleshy pectoral fins. It differs besides from Lepisosteus by the naked body, and from Esox by the tail &c. The name means fleshy arms.

83 Species. OHIO RIBBONFISH. Sarchirus vittatus. Sarchire rubanne.
Back olivaceous brown, and with three longitudinal furrows, a black lateral band from the mouth to the end of the tail, no lateral line. Belly with a lateral row of black dots on each side. Jaws obtuse longer than the head. Anal and dorsal fins ovate acute with two transverse black bands, the anal with ten rays, the dorsal with nine. Tail unequilateral acuminate.

Sarchirus vittatus. Raf. in Journ. Ac. Nat. Sc. Philadelphia. V. 1, page 418, tab. 17. fig. 2.
In the lower parts of the Ohio and at the falls; length from

six to twelve inches. Vulgar names Ribbonfish and Garfish. Not used as food. Abdominal fins narrow almost linear acute, and with two transverse black bands, situated half way between the pectoral and anal fins. This last far from the tail.

XXVII Genus. PIKE. Esox. Brochet.

Body cylindrical or very long covered with small scales, vent posterior. One dorsal fin behind the abdominal fins. Mouth large, jaws long and flattened with very strong teeth: opening of the gills very large. Head bony scaleless. Tail not oblequal. All the fins with rays.

There are several species of Pikes in the Ohio, Mississippi, Wabash, Kentucky, &c. I have not yet been able to observe them thoroughly. I have however procured correct accounts, and figures of two species; but there are more. They appear to belong to a peculiar subgenus distinguished by a long dorsal fin, a forked tail, and the abdominal fins anterior, being removed from the vent. It may be called *Picorellus*. The French settlers of the Wabash and Missouri call them *Piconeau*, and the American settlers Pikes or Pickerels. They are permanent but rare fishes, retiring however in deep waters in winter. They prefer the large streams, are very voracious, and grow to a large size. They prey on all the other fishes except the Garfishes, &c. They are easily taken with the hook, and afford a very good food, having a delicate flesh.

84th Species. STREAKED PIKE. *Esox vittatus*. Brochet raye.

White, with two blackish longitudinal streaks on each side, back brownish: jaws nearly equal, very obtuse, eyes large and behind the mouth: dorsal fins longitudinal between the abdominal and anal fins. tail forked.

E. vittatus. Raf. in American Monthly Magazine, 1818 Volume 3, page 447.

This fish is rare in the Ohio, (although it has been seen at Pittsburgh,) but more common in the Wabash and Upper Mississippi. It is called *Piconeau* or *Picaneau* by the Canadians and Missourians. It reaches the length of from three to five feet. The pectoral and abdominal fins are trapezoidal, the anal and dorsal longitudinal with many rays and nearly equal. It is sometimes called Jack or Jackfish. Lateral line straight.

85th Species. SALMON PIKE. *Esox salmoneus.* Brochet saumonne.

White, with many narrow transversal brown bands, somewhat curved: jaws nearly equal, very obtuse: dorsal fins brown longitudinal and extending over the anal fins: tail forked and brown.

It is one of the best fishes in the Ohio, its flesh is very delicate, and divides easily, as in Salmon, into large plates as white as snow. It is called Salmon Pike, White Pike, White Jack or White Pickerel, and *Picaneau blanc* by the Missourians. It has a short and thick head, eyes not very large, and situated upwards. Pectoral and abdominal fins trapezoidal. Dorsal fin beginning behind these last and extending over the anal. The number of transversal bands is twelve or more, rather distant and with the concavity towards the head. It reaches the length of five feet. Lateral line nearly straight.

XXVIII. Genus. GARFISH. LEPISOSTEUS. Lepisoste

Body cylindrical or fusiform, covered with hard bony scales, vent posterior. Head bony scaleless. Jaws very long, and with strong unequal teeth. Opening of the gills very large. Tail obliqual. All the fins with rays. One dorsal fin behind the abdominal fins which are removed from the vent.

The Garfishes or Gars, are easily known from the Pikes by their large and hard scales. This fine genus had been overlooked by Linneus and united with the Pikes. Lacepede was the first to distinguish it; but he has not been able to ascertain nor elucidate its numerous species. He has blended all the North American species under the name of *Lepisosteus gavial,* the type of which was the *Esox osseus* of Linneus, or rather the Alligator fish of Catesby. I find that Dr. Mitchill, in a late publication, describes another species quite new under the obsolete name of *Esox osseus.* I shall describe and distinguish accurately five species living in the Ohio or Mississippi, which must be divided into two subgenera. To this number must be added three other known species. 1. *L. gavial,* the Garfish or Alligator fish of the Southern Atlantic States. 2. *L. spatula* or the Gar of Chili. 3. *L. indicus* or the East Indian

Gar. I suspect however that there are more than ten species
of these fishes in the United States, and many others in South
America, &c. The Gars of the Ohio partake of the inclina-
tions and properties of the Pikes; but they are still more dangerous and voracious. Their flesh may be eaten: but is often
rejected owing to the difficulty of skinning them, the operation
may however be performed by splitting the skin beneath in zig-
zag. Their scales are very singular, they are not embricated
as in all other fishes; but lay over the skin in oblique rows, and
are as hard as bones. They have many other peculiarities in
common which have been stated by Cuvier, or may be collected from the following descriptions.

1 Subgenus. CYLINDROSTEUS.

Body cylindrical, dorsal fin beginning behind the anal fin.
The name means *bony cylinder.*

56th Species. DUCKBILL GARFISH. *Lepisosteus platostomus.* Lepisoste platostome.

Jaws nearly equal, as long as the head, about one ninth of total length, and flattened; body cylindrical olivaceous brown a-
bove; white beneath: fins yellowish; dorsal and anal spotted with
eight rays; abdominal fins with seven rays, still obtuse-obovel
and spotted with brown: lateral line nearly obsolete.

This species is not uncommon in the Ohio, Miami, Scioto,
Wabash, Mississippi, Missouri, Tennessee, Cumberland, &c.
and other tributary streams. It reaches the length of four feet.
It is taken with the seine, the hook, and even with the gig or
harpoon. It is found as far as Pittsburgh and in the Alleghany
River. Its flesh is as good as that of the Streaked Pike; but is
erroneously thought poisonous by some persons. I shall give a
full description of it, which will preclude the necessity of repe-
tition in describing the others:- The individuals which I observed were 26 inches long, the head 5½, the jaws 2¾ inches:
the dimension from the end of the jaws to the abdominal fins
was 12 inches, and to the vent 18. The body was 2 inches
horizontally and 2¼ vertically; nearly cylindrical, but slightly
flattened on the back and belly, with convex sides slightly yellowish: the whole body is covered with hard bony scales, some
what unequal and obliquely rhomboidal, but with the two inner

sides concave and the two outward sides convex, lying in oblique rows, surface smooth and convex. Head scaleless, hard, and bony, eyes behind the base of the jaws, iris large gilt with a brown stripe across, centre or real eyes small and black. Jaws short, broad, flat and obtuse, breadth about one fifth of the length, the upper one putting over the lower one and with four small nostrils at the end, motionless and with three longitudinal furrows. The lower jaw moveable, soft in the middle. Teeth white, unequal, acute, strong, and upon a single row. Tongue bilobed cartilaginous and rough. Branchial with 8 rays, jutting out and gilt. Pectoral fins yellow with 12 rays, situated directly behind the gill covers and elliptical acute. Abdominal fins yellow, obliquely oboval obtuse and with 7 rays. Anal and dorsal fins oval nearly equal and acute, each with 8 rays the anterior of which is serrated, yellowish olivaceous and spotted with brown, the dorsal beginning behind the beginning of the anal. Space between those fins and the tail attenuated. Tail or caudal fin four inches long, oblong oboval, entire obtuse, base obliqual, the lower part decurrent, with twelve rays, the upper one serrated, yellowish olivaceous spotted with small unequal brown spots. Lateral line concealed under the scales, hardly visible outside. This fish bears (together with the following) the names of Gar, Garfish, Alligator Gar, Alligator fish, Jack or Gar Pike, &c. and on the Mississippi the French names of *Brocheteau, Picaneau, Poisson caymon,* &c.

37th Species. WHITE GARFISH. *Lepisosteus Albus.* Lepisoste blanc.

Jaws nearly equal, as long as the head, about one eighth of total length, and very broad; body cylindrical and white, fins olivaceous unspotted, tail obtuse oblong, lateral line obsolete.

This fish resembles very much the foregoing, and has the general shape of a Pike. It is covered all over with white shining obliqual elliptical smooth and convex scales. It reaches the length of six feet, and is often called Garpike or Pike-gar. It is a rare fish in the Ohio. Jaws shorter and broader than in the foregoing, breadth one fourth of the length.

38th Species. OHIO GARFISH. *Lepisosteus oxyurus.* Lepisoste oxyure.

Upper jaw longer, longer than the head, one sixth of total length, flat and narrow: body cylindrical olivaceous brown above, white beneath: dorsal fin with eight rays, anal fin with ten, abdominal with six, lanceolate acute, spotted with black; lateral line straight, but raised upwards at the base.

This is a very distinct species by the shape of the jaws and tail. It is found in the Ohio; but is by no means common. It reaches six feet in length. Its flesh is not very good to eat, rather tough and strong smelling, like that of some strong sturgeons. The individual which I observed was caught at the falls, and was 30 inches in length, with the upper jaw 5 inches long, while the lower jaw was only four inches: the upper one has three furrows and juts over the lower by a thick curved obtuse point with four small openings or nostrils, although there were two other oblong nostrils in obliqual furrows, at the base before the eyes. This does not appear in *I. platostomus*. Lower jaw straight with a membrane between the lateral lines. Teeth unequal straight very sharp and on a single row. Breadth of the jaws one eighth of the length. Iris large and gilt. Head rough nearly square, covered with six broad plates, two of which on each side, and of a fulvous grey colour. Body cylindrical covered with the usual hard scales in oblique rows, but not two scales exactly alike either in shape or size; they are generally elongated obliquely with the two longest lateral sides straight, the upper one concave and the lower one convex, but these is a row of obcordated ones on the back. All the fins fulvous, the pectoral lanceolate acute with 12 rays, the abdominal lanceolate acute and with only 6 rays. Dorsal and anal trapezoidal elongated, serrated by scaly rays anteriorly. Caudal fins with 12 rays, one sixth of total length, covered with a few large black spots, of a lanceolate shape, with an oblique flexuose base decurrent beneath and acute at the end, serrated both upwards and downwards, and serratures extending on the body. Lateral line not obsolete, quite straight, but raised a little upwards at the base.

89th Species. LONGBILL GARFISH. *Lepisosteus longirostris.* Lepisoste longirostre.

Esox osseus. Mitchill in Amer. Monthly Magazine, Vol. 2, page 321.

Upper jaw longer than the lower and the head one fourth of total length and narrow: body cylindrical, dorsal and anal fins with 8 rays, abdominal fins with 6, tail unspotted nearly trun-sate, lateral line obsolete.

I have only seen the head of this fish, which was taken in the Muskingum. It is evidently the same fish described at length by Dr. Mitchill under tho old Linnean name of *Esox osseus* and found in Lake Oneida; although his description is very minute in some respects, he has omitted to mention the colour of the body, shape of the fins, and many other peculiarities. I refer to his description, and shall merely add its most striking dis-crepancies from the former species. Length forty inches, up-per jaw ten inches with two crooked teeth at the end, lower jaw nine inches, teeth of three sizes crowded on the jaws. Scales rhomboidal. Abdominal fins nearly medial. Tail with 12 rays, serrated above and below.

2d Subgenus. ATRACTOSTEUS.

Body fusiform or spindle shaped, dorsal and anal fins quite op-posite. The name means *bony spindle.*

90th Species. ALLIGATOR GARFISH. *Lesisosteus ferox.* Lepisoste ferox.

Jaws nearly equal, as long as the head, about one eighth of total length and broad: body fusiform and brownish; dorsal and anal fins opposite, tail obliqual oval, lateral line obsolete.

This is a formidable fish living in the Mississippi, principally in the lower parts, also in Lake Pontchartrain, the Mobile, Red River, &c. It has been seen sometimes in the lower parts of the Ohio. It reaches the length of eight to twelve feet, and preys upon all other fishes, even Gars and Alligators. Mr. John D. Clifford told me that he saw one of them fight with an alligator five feet long and succeed in devouring him, after cut-ting him in two in its powerful jaws. My description is made from a sketch drawn by Mr. Clifford, and a jaw bone preserv-ed in his Museum. These jaws are from twelve to eighteen inches long, and from four to six inches broad. They are crowd-ed with teeth, unequally set, not two of which are alike in size,

the largest lie towards the end, and have many small ones between them: they are however all of the same structure, im-
planted in sockets and conical, base grey, striated and hollow,
top white smooth, curved and very sharp. The longest meas-
ure one and a half inch, and are three quarters of an inch thick
at the base. The diameter of the body is nearly one sixth of
the total length. The anal and dorsal fins are small and with
few rays. It is called the Alligator fish or Alligator gar, and
by the Louisianians *Poisson Caymqn.* The scales are large,
convex, and rhomboidal.

XXIX. Genus. DIAMOND FISH. LITHOLEPIS. Litholepe.

Body fusiform, covered with hard stony pentaedral scales,
vent nearly medial. Abdominal fin near the vent. One dorsal
fin opposite to the anal. Head bony scaleless protruded anteriorly in a long snout, mouth beneath the head, jaws not elongated, with strong unequal teeth. Opening of the gills very large.
Tail not obliqual. All the fins with rays.

A very singular genus, which comes very near to the last subgenus; but differs by the snout, mouth, tail, scales, &c. It
must belong however to the same family. The name means
Stony scales.

91st Species. DEVIL-JACK DIAMOND-FISH. *Litholepis adamantinus.* Litholepe adamantin.

Snout obtuse as long as the head; head one fourth of total
length; body fusiform blackish: dorsal and anal fins equal and
with many rays: tail bilobed, lateral line obsolete.

Litholepis adamantinus. Raf. in American Monthly Magazine, 1818, Vol. 3, p.447, and in *Journal de Physique et Hist.
Nat.* 70. N. G. d'Animaux, G. 20.

This may be reckoned the wonder of the Ohio. It is only
found as far up as the falls, and probably lives also in the Mississippi. I have seen it, but only at a distance, and have been
shown some of its singular scales. Wonderful stories are re-
lated concerning this fish, but I have principally relied upon
the description and figure given me by Mr. Audubon. Its
length is from 4 to 10 feet. One was caught which weighed
400lbs. It lies sometimes asleep or motionless on the surface
of the water, and may be mistaken for a log or a snag. It is

impossible to take it in any other way than with the seine or a very strong hook, the prongs of the gig cannot pierce the scales which are as hard as flint, and even proof against lead balls! Its flesh is not good to eat. It is a voracious fish: Its vulgar names are Diamond fish, (owing to its scales being cut like diamonds) Devil fish, Jack fish, Garjack, &c. The snout is large, convex above, very obtuse, the eyes small and black, nostrils small round before the eyes, mouth beneath the eyes, transversal with large angular teeth. Pectoral and abdominal fins trapezoidal. Dorsal and anal fins equal longitudinal with many rays. Tail obtusely and regularly bilobed. The whole body covered with large stone scales laying in oblique rows, they are conical, pentagonal, and pentaedral with equal sides, from half an inch to one inch in diameter, brown at first, but becoming of the colour of turtle shell when dry: they strike fire with steel! and are ball proof!

THIRD PART.—APODIAL FISHES.

Having complete gills, with a gill cover and a branchial membrane. No lower or ventral fins.

XXX. Genus. EEL. ANGUILLA. Anguille.

Body scaleless, elongated. Mouth with small teeth. Pectoral fins. Dorsal and anal fins very long and united with the caudal fins. Vent nearly medial. Gill covers bridled.

It is remarkable that there is only this apodial genus of fish, and not a single jugular genus, in the Ohio, while there are so many abdominal and thoracic genera. The Eels of the Ohio of which I have already ascertained four. species belong all to the subgenus *Conger*, having the jaws nearly equal and obtuse. They are permanent, but rare, and reach a large size. They are taken with the hook, seines, &c. They feed on small fishes, shells, and lobsters, and afford a good food.

92d Species. BROADTAIL EEL. *Anguilla laticauda.* Anguille largequeue.

Black above, white beneath, head flattened, jaws nearly equal, the upper somewhat longer, obtuse and broad. Dorsal fin beginning above the pectorals, which are small and oboval: late-

K

ral line beginning before the pectorals; tail large rounded and dilatated.

It is found in the Ohio in deep and muddy bottoms. Length from two to four feet. Forehead sloping, eyes very small. Dorsal fin and tail black. One individual of this species poisoned once slightly a whole family, causing violent colics, which was ascribed to its having been taken in the vitriolic slate rocks of Silver creek near the falls.

93d Species. Black Eel. *Anguilla viverrima*. Anguille noire.

Entirely black, jaws nearly equal, flat and obtuse: dorsal fin beginning above the pectoral. Tail obtuse.

This species is found in the Tennessee, Cumberland, &c. It differs from the foregoing by being totally black, and not having a broad tail. The body is also somewhat rounded. It reaches the same length. Very good to eat.

94th Species. Yellow-belly Eel. *Anguilla xanthomelas*. Anguille xanthomele.

Black above, yellow beneath, jaws nearly equal, flat and obtuse; dorsal fin beginning over the pectorals. Tail obtuse.

This species is also very much like *A. laticauda*; but it has not the broad tail, the body is thicker, the belly yellow and black &c. It is found but seldom as high as Pittsburgh. Length two or three feet.

95th Species. Yellow Eel. *Anguilla lutea*. Anguille jaune.

Body entirely yellowish; back slightly brownish; throat pale; jaws nearly equal, obtuse, dorsal fin beginning behind the pectorals: tail obtuse, marginated with brown.

It is found in the Cumberland, Green River, Licking River, &c. Length commonly two feet, very good to eat. The lateral line begins over the pectorals, while the dorsal fin begins much behind and pretty near the vent.

FOURTH PART.—ATELOSIAN FISHES.

Having incomplete gills, without a gill cover, or a branchial membrane, or without both.

XXXI. Genus. Sturgeon. Accipenser. Eturgeon.
A gill cover without branchial membrane. Body elongated with three or five rows of large bony scales. Abdominal. Vent posterior. One dorsal and one anal fin. Tail obliqual and unequal. Mouth beneath the snout, toothless, retractible; snout bearded by four appendages before the mouth.

A very interesting and extensive genus, inhabiting all the large rivers of the northern hemisphere; many species are anadromic and live in the sea in the winter. There are six species in the Ohio and its branches, which appear very early in the spring, and must therefore winter in the deep waters of the Mississippi. They are all good to eat and are used as food. They are taken with the seines and harpoons. They spawn in the Ohio, &c. Linneus, Lacepede, Shaw, and Schneider knew very few species of this genus. I have proved, in a Monography, that it must contain about 40 species, of which I have ascertained 20. Seven of them belong to the Old Continent; 1. A. sturio, Linneus. 2. A. huso, L. 3. A. ruthenus, L. 4. A. stellatus, L. 5. A. lichtensteini, Schn. 6. A. lutescens, Raf. 7. A. attilus, Raf.; while thirteen are peculiar to North America; 8. A. atlanticus, Raf. (A. sturio, Mitchill.) 9. A. oxyrinchus, Mitchill. 10. A. rubicundus, Lesueur. 11. A. muricatus, Raf. (var. prec. Lesueur.) 12. A. marginatus, Raf. 13. A. brevirostrum, Les. (His three varieties are probably distinct species.) 14. A. hudsonius, Raf.; besides the six following ones.

1st Subgenus. Sturio.
Five rows of scales on the body, one dorsal, two lateral, and two abdominal.

96th Species. Spotted Sturgeon. Accipenser maculosus. Eturgeon tachete.

A. maculosus. Lesueur in Transactions of the American Philosophical Society; New Series vol 1, page 393.

Head one fourth of total length channelled between the eyes, which are oblong, snout elongated obtuse. Body pentagonal olive, with black spots and small asperities: 13 dorsal scales, lateral rows with 35 scales, abdominal rows with 10.

It is found in the Ohio as far as Pittsburgh. Size small, not exceeding two feet. Mouth and pectoral fins large. Scales rugose, radiated, keeled and spinescent behind. Iris yellow oblong. See Lesueur's description.

97th Species. Shovelfish Sturgeon. *Accipenser platoryn-chus.* Eturgeon pelle.

Head one fifth of total length, flattened; snout flat oval, hardly obtuse, rough above, eyes round. Body pentagonal smooth, pale fulvous above, white beneath. Tail elongated mucronate: 16 dorsal scales, lateral rows with 40; abdominal rows with 12. A singular species, very common in the Ohio, Wabash, and Cumberland in the spring and summer; but seldom reaching so high as Pittsburgh. It appears in shoals in March, and disappears in August. It is very good to eat and bears many names, such as Spade-fish, Shovel-fish, Shovel-head, Flat-head, Flatnose, &c. having reference to the shape of its head, which is flattened somewhat like a spade. It is also found in the Mississippi and Missouri, where the French call it *La pelle* or *Poisson pelle*, which has the same meaning. Size from two to three feet, greatest weight 20 lb. Body rather slender, with small bluish dots on the back and whitish on the sides. Dorsal scales brownish, radiated, punctuated, and spinescent. Lateral scales diminiated, serrated behind, the posterior smaller: the abdominal nearly similar, hardly serrated. Two nostrils on each side before the eyes, the posterior larger oblong obliqual. Eyes round black, iris coppered. Mouth with eight lobes and verrucose. Tail very long, one fifth of total length, the upper lobe scaly above, slender and with a long filiform terminal process. All the fins trapezoidal, the dorsal falcated with 26 rays and nearly opposite to the anal. Pectoral large 46 rays. Abdominal 20. Anal 14. Tail, inferior lobe 18, superior 60.

2d Subgenus. Sterletus.

Only three rows of scales, one dorsal and two lateral.

98th Species. Fall Sturgeon. *Accipenser scrotimus.* Eturgeon tardif.

Head conical two ninths of total length, snout short obtuse, eyes somewhat oblong. Body cylindrical entirely fulvous brown, belly white. Tail short and truncate obliquely. Dorsal scales 17, two of which behind the dorsal fin, lateral rows with about 30 scales.

A large species reaching 5 and 6 feet in length. It appears in June and disappears in November, but is seldom caught, except in the fall, when attempting to go down the river. It is sometimes caught in the Kentucky as late as November. It affords a tolerably good food. Snout very short yet somewhat attenuated, barbs brown, eyes nearly round, head with a depression above, lips very thick. Scales radiated knobby behind. Pectoral and anal fin somewhat oboval, the abdominal and dorsal trapezoidal.

99th Species. OHIO STURGEON. *Accipenser ohiensis*. Eturgeon del' Ohio.

Head conical one fifth of total length, snout sloping short nearly acute, eyes round. Body cylindrical rough olivaceous fulvous, belly white. Tail short lunulate falcate. Dorsal scales 14 carinated, the lateral rows with 34 dimidiated and unqual.

Somewhat similar to the foregoing. Length from three to four feet. Found as far as Pittsburgh, comes in the spring, and goes away in September. Head convex above, with a protuberance on the top. All the fins trapezoidal but somewhat falcate. The tail remarkably so, and obliquely lunulate, the lobes not divided by a notch as usual in the other species. It has been mentioned by Lesueur as a variety of his *A. rubicundus*, page 390 of the Trans. Am. Phil. Society, but it differs widely from it.

100th Species. BIGMOUTH STURGEON. *Accipenser macrostomus*. Eturgeon beant.

Head one fourth of total length, snout elongated, somewhat flattened, eyes round. Body cylindrical deep brown above. white beneath. Tail elongated; about 20 dorsal scales, several between the dorsal and anal fin, about 30 scales in each lateral row.

I have not seen this species, but Mr. Audubon has communicated me a drawing of it. It is only found in the lower parts of the Ohio, and reaches four feet in length. Good food, Mouth large gaping, hanging down, retractible. Gill cover oblong. Tail slender, the lower lobe very small. Fins trape-

zoidal, the dorsal and anal somewhat falcated and more distant
from the tail than usual.   Lateral scales dimidiated.

XXXII Genus.   DOUBLE FIN.   DINECTUS.   Dinecte.

Differs from Sturgeon, by having two dorsal and no abdominal fins.   First dorsal anterior, the second opposed to the anah
Three rows of scales as in *Sterletus.*

This genus rests altogether upon the authority of Mr. Audubon, who has presented me a drawing of the only species belonging to it.   It appears very distinct if his drawing be correct; but it requires to be examined again.   Is it only a Sturgeon incorrectly drawn?

101st Species.   FLATNOSE DOUBLEFIN.   *Dinectus truncatus.*
Dinecte camus.

Head one fifth of total length, conical, snout very short, truncated, eyes round.   Body cylindrical deep brown above, silvery white beneath, tail elongated: dorsal scales, 4 before the first dorsal fin, 6 between the fins, and 4 behind the second, lateral rows with about 30 small dimidiated scales.

This fish was taken with the seine near Hendersonville in the spring of 1818 by Mr. Audubon.   Length two feet, skin very thick and leathery.   Mouth very large and hanging down as in the foregoing, somewhat like a proboscis.   Bones land anal fins trapezoidal, dorsal fins nearly triangular, the first larger and standing immediately behind the pectoral.   Gill cover rounded.   Tail somewhat forked, the upper lobe thrice as long as the lower.   Four long white barbs, very near the end of the snout, eyes above the mouth.

XXXIII Genus.   SPADEFISH.   POLYODON.   Polyodon.

Differs from Sturgeon, by having a transversal mouth with teeth, no barbs and no scales.   Snout protruded in a long flat process, gill cover elongated by a membraceous appendage.

This singular genus was first described by Lacepede.   It belongs to the family of *Sturionia,* along with the two foregoing and the following.   Only one species is known as yet.

102d Species.   WESTERN SPADEFISH.   *Polyodon folium.*
Polyodon feuille.

Head longer than the body, snout as long as the head, cunei-

form obtuse thin and veined with one main nerve. **Brown a**-
bove, white beneath.

Squalus spathula Lacep. Poiss. 1, p. 403, tab. 12, fig. 3.
Polyodon folium Lacep. and Auct. mod.
Spatularia. Schneider's Ichthyology.

This singular fish has often been described and figured, but
I have not seen a single figure of it perfectly correct. It is a
rare fish, occasionally seen in the Mississippi, Missouri, Ohio,
&c. It disappears in winter. I saw several at the falls in September 1818. It is caught in the scines and sometimes bites
at the hook. It is not eaten. Length from one to three feet. I
shall add an exact description of it. An oblong redish spot at
the base of the snout, which is brown membranaceous, with a
thick cartilaginous nerve in the middle and many veins; broader
and obtuse at the end. Eyes round small black, before the
mouth, a small nostril in front of them. Mouth large, similar
to that of a shark, with small crowded teeth on the jaws and
the tongue, this is large thick and similar to a file. Gill cover
very long membranaceous reaching the abdominal fins. A lateral line following the curve of the back. All the fins brown,
nearly rhomboidal, with an obliqua redish band, and a multi-
tude of small crowded rays, inserted on a thick fleshy
lump: the dorsal fin larger and rather more anterior than the
anal. Tail very obliqual, serrated above: lobes not very different in size, but extremely in shape and situation, the lower one
broader, shorter, and nearly triangular.

XXXIV Genus. PADDLEFISH. PLANIROSTRA. Planirostre.
Differs from *Polyodon*, by having no teeth whatever and the
gill-cover radiated with a short appendage.

By the want of teeth this genus is intermediate between *Po-*
lyodon and *Accipenser.* It was first described by Lesueur, under the name of *Platirostra* (by mistake) instead of *Planirostre:*
I had called it in manuscript *Megarhinus paradoxus.*

103d Species. TOOTHLESS PADDLEFISH. *Planinostra eden-*
tula. Planirostre edente.

Head as long as the body, snout longer than the head, somewhat cuneiform, obtuse, and thin, with two longitudinal nerves

and reticulated veins forming an hexagonal network. Body entirely olive brown.

Platirostra edentula, Lesueur in Journ. Ac. Nat. Sc. Philadelphia, Volume 1, page 229.

This fish is still more rare than the foregoing, but found occasionally as far as Pittsburgh. It is larger, reaching from 3 to 5 feet and 50lbs weight. Not very good to eat. It has been so fully described by Lesueur, that I need not do it again. The individual which I saw was 40 inches long, head 20 inches, snout 11 inches long and 2½ wide at the end, hardly cuniform. Eyes exceedingly small and round. Gill cover oval radiated as in the Sturgeons, with a short membranaceous flap, reaching only beyond the pectoral fins, &c. It is also called, along with the foregoing, Oar fish and Spatula fish.

XXXV Genus. LAMPREY, PETROMYZON. Lamproie.

Body cylindrical scaleless, vent posterior. Two dorsal fins and a caudal fin, no other fins. Seven branchial round holes on each side of the neck. Mouth terminal inferior acutiform, toothed.

There are two or three species of Lampreys in the Ohio, but they are very scarce and I have only seen one as yet.

104th Species. BLACK LAMPREY. *Petromyzon nigrum*. Lamproie noire.

Entirely blackish, tail oval acute, second dorsal over the vent, several rows of teeth.

A very small species, from four to five inches long; it is found as high as Pittsburgh. Dorsal fins shallow, and distant from each other and the tail. Eyes round and large. Branchial holes small. No lateral line. Mouth oval, teeth white and yellow. It torments sometimes the Buffaloe fish and Sturgeons, upon which it fastens itself. It is never found in sufficient quantity to be used as food.

SUPPLEMENT.

THE Itchtfyology of the River Ohio was begun to be printed in the Weatern Review in December 1819, and has been continued gradually unil November 1820. During the course of the impression some new species have been discovered, or ascertained, which I now propose to notice

THORACIC FISHES.

XXXVI Genus. SPRINGFISH PEGEDICTIS. Pegedicte.
· Body conical with small scales, belly flat, vent medial. Head broad scaleless, gill cover with a membranaceous appendage and a concealed spine, mouth toothed. Two dorsal fins, the first with simple, soft, semi-spinescent rays. Thoracic fins with five rays.

This new genus belongs to the family Percidia, and has many affinities with the G. Holocentrus Lepomis, Etheostoma, &c. but its conical form and many other secondary peculiarities distinguish it completely. The name means Fountain-fish.

105th Species. CATSEYE SPRINGFISH. Pegedictis ictalops. Pegedicte œuil de chat.
Jaws equal, forehead knobby, eyes elliptical. Body olivaceous with some black transversal unequal brown bands; a concealed spine on the gill cover: lateral line straight: tail elliptical. The first dorsal fin with 8 rays, the second with 12, as well as the anal and pectoral fins.

I have discovered this species in the summer of 1820 near Lexington. It has no vulgar name. Length hardly two inches. Head large brown, convex above with several small knobs on the forehead, flat beneath. Eyes as in the Catfishes with oblong eyes, iris gilt brown. Spine of the gill cover concealed under the skin. Teeth small and acute. Pectoral fins large lanceolate. Belly white and flat. Fins hyalin with some brown spots. Five transversal bands. The specific name means Catseye.

6th Genus. ETHEOSTOMA.

106th Species. SPRINGS HOGFISH. Etheostoma fontinalis. Etheostome des fontaines.
Body oblong cylindrical, breadth one sixth of the length, olivaceous, sides with transversal brown lines somewhat curved: a small round black spot behind the gill cover; lateral line obsolete. Jaws obtuse, the upper one shorter. Tail oboval entire gilt tesselated with black. First dorsal with 8 rays, the second and anal with 12.

A little species, from one to two inches long, found in the springs and caves near Lexington in the summer. It belongs to the subgenus Diplesion. Body cylindrical somewhat compressed. Head small flat above: gill cover attenuated behind

F

obtuse and with a spine. Eyes small, iris gilt. Dorsal fins joining, the first with spiny rays appendiculated, second with soft rays, anal fin opposed to it and with two spiny rays. Pectoral lanceolate with 12 rays, thoracic lanceolate with 6. Vent anterior.

ABDOMINAL FISHES.

17th Genus. SEMOTILUS.

107th Species. SILVERSPOTTED CHUBBY. *Semotilus? notatus.* Semotile tache.

Breadth one sixth of the length, brownish, pale beneath; head small obtuse with a large silver spot on the forehead before the eyes, ~~jaw~~ ~~~~ ~~~~ ~~~~ to the anal, tail oboval ~~~~

~~~~ Cumberland River, and the Little River, a branch of it. Communicated by Mr. Wilkins. It is rather doubtful whether it belongs to this genus, or *Minnilus, Rutilus,* &c. It might perhaps be found to constitute a peculiar one by the small mouth without lips, and the posterior dorsal fin. Vent posterior. Pectoral and abdominal fins oboval. Eyes large. Length three inches, good bait for Perch, Bass, Redeyes or Ringeyes, &c.

26th Genus. SARCHIRUS.

108th Species. SILVER RIBBONFISH. *Sarchirus? argenteus.* Sarchire argente.

Entirely silvery, without bands or spots.

Communicated by Mr. Owings. It is found in Licking River, ~~Slate Creek, &c.~~ Length from two to three feet. It is called Pike and may be one, but as it is described without scales and very slender, I have added it to this genus, until it is better known.

~~ORIAN FISHES.~~

~~~~ ~~~~ *agenarius.* Esturgeon gourde.

Snout attenuated obtuse like a gourd, body entirely brown.

A species of Sturgeon which I have never seen, is said to live in the Ohio, which is called Gourdfish owing to its head having the shape of a gourd, of which the snout represents the neck. It reaches two and three feet in length.

XXXVII Genus. SAWFISH. PRISTIS. Poisson-Scie.

Abdominal, with five branchial spiracles on each side, body cylindrical, tail obliqual, head protruded in a long saw.

This genus belongs to the family of Sharks or *Antacea.*

110th Species. MISSISSIPPI SAWFISH. *Pristis Mississippiensis.* Poisson-Scie du Mississippi.

Saw thicker in the middle where it has two longitudinal furrows; margin somewhat sinuated with transversal depressions, 26 long and narrow acute teeth on each side, alternating with the depressions: extremity of the saw rounded nearly truncate, with a raised granular margin reflected upwards.

I have only seen the saw of this fish, which is preserved in Mr. Clifford's museum. It is six and a half inches long, and one broad, olivaceous above, pale beneath, middle part raised but flat. Teeth half an inch long, shorter and more distant near the base, 26 on the right and 27 on the left, nearly equal. This fish is found in the Mississippi, Lake Pohtchartrain, Red River, Arkansas, Mobile, and has even been seen in the Ohio, length from three to six feet.

XXXVIII Genus. HORNFISH. PROCEROS. Proceros. Apodal. Body elongated. Vent posterior. One dorsal fin opposed to the anal. Mouth beneath transversal toothed Snout protruded in a a straight horn. Four spiracles or branchias on each side.

Singular new genus of the family of Sharks or *Antacea*, from which however it differs by the want of abdominal fins. There are two species of it: the second, which I have called *Proceros vittatus*, lives in Lake Ontario, and has longitudinal stripes.

111th Species. SPOTTED HORNFISH. *Proceros maculatus.* Proceros tachete.

Iron grey with white spots on the sides: tail forked: horn one fourth of total length.

This fish lives in the Mississippi, and is sometimes caught at St. Genevieve in the State of Missouri. The French settlers call it *Poisson arme*. It has no scales, but its head is bony: Eyes very small. Dorsal and anal fins rounded. Length two or three feet, very good to eat. Communicated by Mr. M—— of St. Genevieve.

Several imperfect and incorrect notices or Catalogues of fishes living in the western waters have been published. Carver and Pike have noticed those of the Upper Mississippi, Curtis those of Red River, Pike those of the Arkansas and Osage rivers, Thomas those of the Wabash, and Lewis and Clarke those of the Missouri; but very few practical facts can be collected from their imperfect accounts, except perhaps from the two latter travellers. I may at a future period notice the new fishes of the Missouri, discovered by Lewis and Clarke. I shall at present merely add some facts lately ascertained or drawn from Thomas's account of the fishes of the River Wabash, page 211 of his travels published in 1819.

2d Sp. *Perca chrysops*, is found in the Wabash, and called Rock-mullet, it reaches three feet in length and fifteen prounds in weight. This fish will not bite at the hook, unless when it is withdrawn, it then darts on it.

4th Sp. *Amblodon grunniens*, It is sometimes called *Drum* in the Wabash.

14th Sp. *Lepomis flexuolaris*... Mr. Wilkins has informed me that this fish watches over its spawn, and prevents any small fish from coming near it: while thus employed it will not bite at the hook, but endeavours to drive away the bait. It is common in all the tributary streams of the Ohio, also in the Arkansas Osage, Missouri, &c.

19th Sp. *Aplocentrus calliops*, Found in the Cumberland Tennessee, Little River, &c. and called Redeyes or Ringeyes.

63d Sp. *Catostomus erythrurus*, In the Wabash, weighing as far as 15 pounds.

71st Sp. *Pimelodus ~~vidus~~*. It is called *Wal-hea* or Deep w~~~~. The other Catfishes are generally called *Wi-sa-meek* by the same Indians, which means Fat fish. The names of Pout and Bullheads are given to some species in the Wabash, Miami, Mississippi, &c. The French settlers call them Barbottes.

84th Sp. *Esox vittatus*, Thomas mentions three kinds of Pikes found in the Wabash, 1, River Pike; 2, Pond Pike, slim, three feet long, excellent. 3, Jack Pike or Pickerel, excellent, w~~~~.

89th Species. *Lepisosteus longirostris*. ~~~~ the Wabash, called Gar or Billfish, ~~~~ long and quite slim; bill six inches and pointed. It is a strong fish. Thomas says that, having caught them in his hands, he was unable to hold them.

CORRECTIONS AND ADDITIONS.

rection~~~~

Page 15, line 5, Pittsburgh had only 8000 inhabitants by the census of 1820, and Cincinnati about 9000.

Page 19, l. 22. The Cumberland has a fine fall in Kentucky near Monticello.

Page 21. *Perca salmonea* add Raf. 1818 in Amer. Month. Mag. V. 3, p.354

Page 29, l. 1, *Engikreps* read *Exythrops*.

Page 34, l. 17 add *Bodianus calliops*, Raf. 1818, Am M Mag V 3, p 457.

Page 38, l. 4, add *Sciena caprodes*, Raf. 1818 in Am M Mag 3, p 394.

Page 40, l. 2, add *Clupea heterurus*, Raf. 1818, in Am M Mag 3, 355.

Page 42, l. 18, add *Glossodon heterurus*, Raf. in Am M Mag 3, p 354.

Page 43, l. 2, add *Glossodon harengoides*, Raf. in Am M Mag 3, p 354.

Page 43, l. 35, *Hyodon Clodalus* read *Hyodon tergisus*.

Page 48, l. 28, *Luxilus* read *Alburnus*.

Page 49, l. 35, *Semotilus* read *Semotilus*.

Page 50, l. 10, *Diplemia* read *Diplemiss*.

Page 52, l. 27, *Flat-head* read *Fat-head*.

Page 55, l. 21, add Raf. 1818, in American Monthly Mag. 3, p 355.

Page ~~~~, l. 5, add Raf. 1818 in American Monthly Magazine, 3 p 355.

Page ~~~~ 21 add *Silurus punctatus*, Raf. 1818 in Am M ~~~~

Page 64, l. 36, add *Silurus olivaris*, Raf 1818 in Am M Mag 3, p ~~~~

Page 77, add to *Anguilla laticauda*, Raf 1818 in Am M Mag 3, p 447.

INDEX

Luxilus - - Genus 16.	Pike - - -	G. 27.
Minnilus - - - 15.	*Poisson arme* -	G. 38.
Moxostoma - - - 20.	*Poisson cayman* -	28.
Nemocampsis - - 5.	*Poisson lunette.* - -	20.
Notemigonus - - 12.	Pucker -	22.
Noturus - - - 24.	Redbelly -	Sp. 11.
Pegedictis - - - 36.	Redeyes - -	9, 19
*Perca - - - - 1.	Redfish - -	50, 51.
*Petromyzon - - 35.	Red horse ⟩	63.
*Pimelodus - - - 22.	Red tail ⟨	
Pimephales - - 19.	Ribbonfish - -	G. 26.
*Planirostra - - - 34.	Salmon - - -	Sp. 1.
Pogostoma - - 8.	Sawfish -	G. 37.
Polyodon - - - 33.	Shad - -	Sp. 26, 27.
Pomolobus - - 10.	Shiner - -	G. 16.
Pomotis - - - - 4.	Skimback -	Sp. 43, 58.
Pomoxis - - - 6.	Silverfish - -	46, &c.
*Pristis - - - - 37.	Shovelfish - -	G. 33.
Proceros - - - 38.	Springsfish - -	36.
Pylodictis - - 23.	Sturgeon - -	84.
Rutilus - - - 18.	Sucker -	20
*Salmo - - - 14.	Suckrel - -	21.
Sarchirus - - - 26.	Sunfish Sp. 6 to 12, 20.
Semotilus - - - 17.	Toadfish -	60.
*Sciena - - - 2.	Toter - -	G. 25.
Sterletus - - - 31.	Trout - Sp. 15, 34, 35.
Sherodictus - - 31.		20.
Sturio - - - 31.		
Telipomis - - 4.		
Teretulus - - 20		

N. B. The names with asterisks are old generic names: those in italic are new subgenera, or French names in the second column.

www.ingramcontent.com/pod-product-compliance
Lightning Source LLC
Chambersburg PA
CBHW021953190326
41519CB00009B/1240